Music, Sound,
and Technology

Music, Sound, and Technology

John M. Eargle

VNR VAN NOSTRAND REINHOLD
New York

Library of Congress Catalog Number 89-70491
ISBN 0-442-31851-0

Printed in the United States of America

Van Nostrand Reinhold
115 Fifth Avenue
New York, New York 10003

Chapman & Hall
2-6 Boundary Row
London SE1 8HN, England

Thomas Nelson Australia
102 Dodds Street
South Melbourne, Victoria 3205, Australia

Nelson Canada
1120 Birchmount Road
Scarborough, Ontario M1K 5G4, Canada

16 15 14 13 12 11 10 9 8 7 6 5 4 3 2

Library of Congress Cataloging-in-Publication Data

Eargle, John.
    Music, sound, and technology / John M. Eargle.
       p. cm.
    ISBN 0-442-31851-0
    1. Music--Acoustics and physics. I. Title.
ML3805.E2   1990
781.2'3--dc20
                                           89-70491
                                            CIP
                                           MN

Contents

Preface

This book is about music, the instruments and players who produce it, and the technologies that support it. Although much modern music is produced by electronic means, its underlying basis is still traditional acoustical sound production, and that broad topic provides the basis for this book.

There are many fine books available that treat musical acoustics largely from the physical point of view. The approach taken here is to present only the fundamentals of musical physics, while giving special emphasis to the relation between instrument and player and stressing the characteristics of instruments that are of special concern to engineers and technicians involved in the fields of recording, sound reinforcement, and broadcasting.

In order to understand musical instruments in their normal performance environments, the student must have a basic working knowledge of physical and architectural acoustics. The book begins with a review of the elements of acoustics, stressing the nature of sound fields and phenomena that are wavelength-dependent. The book then moves on to a discussion of those aspects of psychological acoustics that are of special concern to music technicians, most notably concepts of stereophonic imaging, loudness-related phenomena, and critical band theory.

Musical scales and their evolution are then discussed, largely as a summary of the problems of earlier centuries and their specific solutions. (We note throughout the book that many of the remaining problems of temperaments and tuning are sufficiently "hidden" behind many modern musical textures.)

We then turn to the basic principles of sound production, discussing vibrating strings, air columns, and vibrating bars and membranes, along with the methods of actuating them. At this point, the discussion does not address specific musical instruments.

Then follows a series of five chapters that discuss, in order, string instruments, woodwind instruments, brass instruments, percussion instruments, and keyboard instruments. These chapters present the historical evolution of the several groups, placing them in the context of musical ensembles. Playing techniques are discussed in detail, and pertinent measurements, such as dynamic range, directional properties, and frequency spectrum, are given. The intent in these chapters is to give the technician, who may not have studied music, the necessary familiarity with and understanding of the instrumental groups to enable him to place his microphones wisely and not make unreasonable demands of the players.

Musical ensembles are then discussed in terms of historical evolution, stage seating, and problems of balance between instrumental sections. Following this is a discussion of performance spaces for music and speech. The unanswered questions of concert hall design are presented, along with some of the newer quantitative measurement and estimation methods. Speech intelligibility has always been easier to define and measure than "musical intelligibility," and the methods of doing so are presented. The problems of multipurpose halls are covered, and some of the possible solutions via "the electronic hall" are offered.

The technologies of speech and music reinforcement are then presented in terms of the requirements for adequate signal level, system gain and stability, and audience coverage in a variety of environments. Sound recording is discussed in terms of both its tools (microphones, signal routing and processing, recorders) and its techniques and esthetic dimensions (microphone placement, creating the desired sound stage).

The final three chapters give an overview of the fields of home high fidelity sound reproduction, musical tone synthesis, and electronic (active) noise control. While high fidelity sound and music synthesis are mature technologies, active noise control is now coming into its own and promises to make a considerable contribution in a crowded world where both noise and music may have gotten somewhat out of hand.

The book is best studied from the beginning, and the overall organization is as follows:

1. Chapters 1–3: Tutorial, covering physical acoustics, psychological acoustics, and temperaments

2. Chapter 4: Basics of musical sound generation

3. Chapters 5–9: Specifics of the major instrumental groups

4. Chapter 10: Musical ensembles

5. Chapter 11: Performance spaces

6. Chapter 12: Speech and music reinforcement

7. Chapter 13: Sound recording

8. Chapters 14–16: Overviews of home high fidelity, music synthesis, and active noise control

The book is ideal as a single-semester text in general music or communications programs at the collegiate level, and it also addresses many of the problems faced daily by recording engineers and other sound professionals.

The author acknowledges the many sources of graphic material used in the book. Credits are presented with the illustrations. The fundamental work of Olson, Benade, Meyer, and others cited here is also gratefully acknowledged.

1

Fundamental Mathematical and Physical Concepts in Acoustics

The student of musical acoustics should have an intuitive feel for basic physics and mathematics. Freshman-level courses in elementary mechanics and algebra are useful prerequisites, but are not absolutely essential. What is required is familiarity with the basic physical quantities of length, mass, and time, and the related concepts of velocity, acceleration, power, and energy. On these fundamentals of mechanics we will construct models of resonant systems and subsequent propagation of sound through various media. Basic sound fields will be discussed, as will the concepts of level and the decibel.

Then we will examine, largely in a qualitative way, the directional properties of sound radiators, moving on to a broad discussion of environmental effects and interference effects.

1.1   THE BASIC QUANTITIES: LENGTH, MASS, AND TIME

Length is the physical measure of extent or expanse. A city block may be 200 meters long; a field may have an area of 200 square meters (m^2). The volume of a large vat may be 200 cubic meters (m^3). Such terms as "distance traveled," "displacement," "extent of space," or simply "size" may be related to convenient units of length, breadth, and width. The unit of length that we will use in this book is the meter, and the unit symbol is m.

The term *mass* refers to the amount of matter or substance, as opposed to its extent. A container of air has less mass than the same container filled with water. Most of us associate mass with weight, since that is our intuitive sense of it. But a given mass of material will have no weight in outer space. The real measure of mass is its *inertia*, which is its tendency to remain motionless, or move along a straight line, unless acted upon by some external force. The unit of mass is the kilogram, and the unit symbol is kg.

Time, of course, is the measure of intervals in our life experience. The second is the fundamental unit, and its symbol is sec.

1.2   VELOCITY AND ACCELERATION

An object moving in a straight line has a velocity, which we can measure as displacement per unit time, or meters per second. If the object has a velocity of 2 m/sec, it will, after 10 seconds, have traversed a distance of 20 meters. The term *speed* is the everyday equivalent.

The following equation describes the concept:

$$\text{Velocity} = \frac{\text{displacement}}{\text{time}} \qquad (1.1)$$

Let us assume that the object is, for whatever reason, moving faster and faster. We then say that it is accelerating, or that it is "speeding up." *Acceleration* can be described as the rate of change of velocity with time. Positive acceleration indicates an increase in velocity, while negative acceleration indicates a decrease in velocity.

Acceleration is measured in meters per second^2, or meters per second per second. What this states is that the velocity is increasing as time elapses. As an example, let us assume that an object has an acceleration of 2 m/sec^2. Starting from rest, after 1 sec its velocity will be 2 m/sec, after 2 sec the velocity will be 4 m/sec, after 3 sec the velocity will be 6 m/sec, and so forth.

The following equation describes the concept:

$$\text{Acceleration} = \frac{\text{velocity change}}{\text{time}} \qquad (1.2)$$

1.3   VECTORS

A *vector* is a quantity that has both magnitude and direction. Both velocity and acceleration are vectors, since any moving object can be described by the magnitude of its velocity and acceleration along with its direction. Mass, length, and time, on the other hand, are not vector quantities, since there is no direction intrinsically associated with them. Such quantities as these are called *scalars*.

1.4   FORCE, ENERGY, AND POWER

If we push an object, we are applying *force* to it. The magnitude of the force can be measured as the product of the mass of the object and the acceleration we are imparting to it. Stated another way, if we apply a constant force in a given direction to an object, it will move in that direction at some constant acceleration. The unit of force is the newton, and its symbol is N.

The following equation describes the concept:

$$\text{Force} = \text{mass} \times \text{acceleration} \qquad (1.3)$$

Gravity acts as acceleration, imparting a downward force, called weight, to any mass within its field. The acceleration due to gravity is 9.8 m/sec^2. What this states is that an object under free fall due to gravity will have a velocity after 1 sec of 9.8 m/sec, a velocity after 2 sec of 19.6 m/sec, and so forth.

Pressure is a measure of how concentrated a force is. The unit is the pascal, which is equal to one newton per square meter. The symbol for pascal is Pa. If a given force is applied over a very small area, the resulting pressure will be large. The same force may be applied over a large area with correspondingly smaller pressure.

Energy and *work* are equivalent in terms of fundamental units, but their physical descriptions are somewhat different. We usually think of energy as the capability of doing work, or that energy is "converted" into useful work. Work may be defined as the application of a given force acting over a given distance. The unit is the joule, indicated by the symbol J. For example, if we move an object over a distance of 10 m by applying a constant force of 1 N to it, we have done work on the object, or expended energy of 10 J.

We can speak of several kinds of energy. If we raise a mass of 1 kg a distance of 10 m above a reference plane, the suspended mass will have a *potential energy* (PE) of 10 J with respect to the reference plane. That is, it has the potential of doing 10 J of work as it falls back to the reference plane. Likewise, a spring under the force of compression or tension will have a potential energy as it exerts its force over a given distance upon being released. The potential energy in the loaded spring is

$$PE = 1/2kx^2. \tag{1.4}$$

where k is the spring constant and x is the displacement of compression in meters. The spring constant is a measure of how strong the spring is, and its dimensions are force per unit displacement, or N/m.

A mass in motion has what is called *kinetic energy* (KE). For example, when a large moving object hits a wall, coming to a halt, it "does work" on the wall by knocking a hole in it. The kinetic energy in a moving mass is

$$KE = 1/2mv^2. \tag{1.5}$$

where m is the mass in kilograms and v the velocity in meters per second.
(Equations 1.4 and 1.5 are presented here without derivation. They come from elementary calculus as applied to mechanics.)

Heat is a form of energy, but it belongs in the realm of thermodynamics rather than mechanics. It will not be necessary to consider heat as we develop our models of simple vibrating and sound radiating systems.

Power is the rate at which work is done. For example, it takes a certain amount of work to move a given object a given distance. If we apply a low-power engine to the task, we may take all day to accomplish it. A more powerful engine can do the job in a shorter period of time. The unit of power is the watt, and it has the dimensions of joules per second. The symbol is W.

1.5 A SIMPLE OSCILLATING SYSTEM

Figure 1-1a shows a mass hanging on a spring under the static force of gravitational acceleration. As we have all experienced, a simple displacement of the mass up or down will result in an oscillatory bobbing motion. Eventually the motion dies out, but in the process it may execute many cycles of up-and-own motion.

In figure 1-1*b*, the mass has been displaced upward a distance *x*. Since the spring has been compressed, it has a potential energy of $1/2kx^2$. When the mass is released, it moves downward, eventually reaching its original static position, but still very much in motion, as shown in figure 1-1*c*. At that point its velocity is at a maximum, and it has kinetic energy of $1/2mv^2$.

Of course the mass overshoots the mark, eventually comes to a halt, and begins its return trip upward. The motion is called *oscillation*, and it exhibits the alternate transfer of energy from potential to kinetic and back again until the motion dies out.

We all know intuitively that the greater the mass, the slower the oscillation, or the stronger the spring, the faster the oscillation. By making precise measurements of the mass and the spring constant, we could arrive at an empirical equation that would tell us beforehand what the frequency of oscillation would be. However, a more elegant solution to the problem comes from the mathematics of differential equations:

$$\text{Frequency} = \left(\frac{1}{2\pi}\right)\sqrt{\frac{k}{m}} \tag{1.6}$$

Frequency is measured in cycles per second and has the dimensions of 1/sec. The term hertz (symbol Hz) is universally used today as the measure of cycles per second.

1.5.1   Other Resonant Systems

There are other resonant systems, as shown in figure 1-2, and they are all governed by the same differential equation that gave us equation 1-6. A pendulum (figure 1-2*a*) has a frequency of resonance given by

$$\text{Frequency} = \left(\frac{1}{2\pi}\right)\sqrt{\frac{g}{l}} \tag{1.7}$$

where *l* is the length of the pendulum in meters and *g* is the acceleration of gravity. Obviously, the longer the pendulum, the slower the frequency of oscillation. We cannot change the acceleration of gravity, but it seems clear that a pendulum of given length would swing more slowly on the moon than on earth because of the lesser gravitational pull of the moon.

A torsional oscillating system is shown in figure 1-2*b*. Its equation is

$$\text{Frequency} = \left(\frac{1}{2\pi}\right)\sqrt{\frac{K}{I}} \tag{1.8}$$

where K is the rotational spring constant and I is the rotational moment of inertia. We will not define these terms here; our purpose is only to demonstrate that equations 1.7 and 1.8 are of the same form as equation 1.6.

1.5.2   Simple Harmonic Motion

The motion described by the various oscillating systems we have discussed can be viewed as the projection of a point on a circle as that point moves around the circle with fixed velocity, as shown in figure 1-3. The projection outlines what is called a sine wave, and the period of the

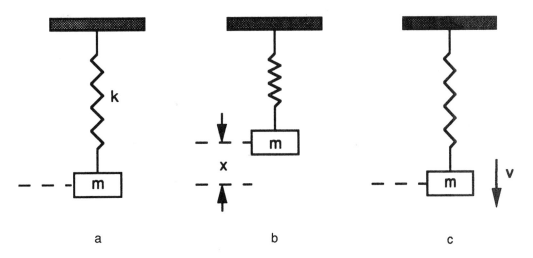

Figure 1-1. A mass suspended by a spring. (*a*) Mass at rest; (*b*) mass displaced by distance *x*; (*c*) mass returned to initial position with downward kinetic energy.

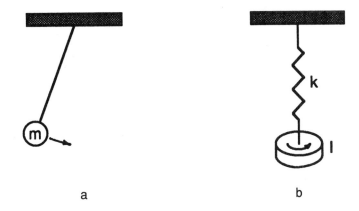

Figure 1-2. Other mechanical resonant systems. (*a*) A pendulum resonator; (*b*) a torsional resonator.

sine wave, T, is the time in seconds required for one complete cycle. Amplitude describes the extent of the displacement, and phase [indicated by the Greek letter phi (ϕ)] describes the fixed time relationship between two sine waves of the same frequency.

An important equation is:

$$f = \frac{1}{T} \quad \text{Hz} \tag{1.9}$$

The frequency *f* of oscillation is the reciprocal of the period.

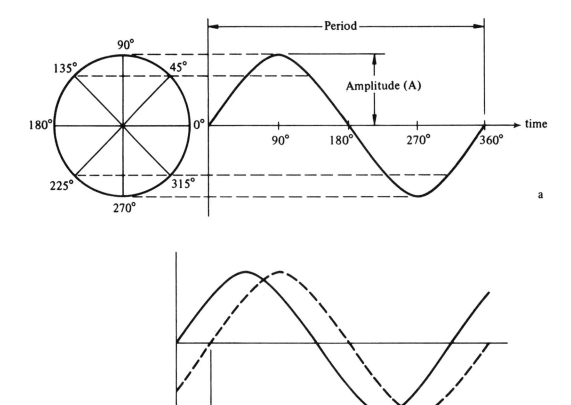

Figure 1-3. Simple harmonic motion shown graphically. (a) Generation of a sine wave of amplitude A; (b) illustration of phase angle ϕ.

1.5.3   Damping

Why do these simple oscillating systems eventually stop? Without a continuing source of energy, these systems will eventually run down because of mechanical losses and air losses. An equation of motion for any damped oscillating system is given:

$$\text{Amplitude} = Ae^{-t/T} \sin(2\pi t) \tag{1.10}$$

Here, the exponential coefficient describes the decay of oscillation with time and its eventual stopping, as shown in figure 1-4. The time constant T, stated in seconds, is a measure of how long it takes for the system's oscillations to lose 63% of their initial amplitude. A is the initial amplitude of the oscillation, and t is the time variable measured in seconds. The more rapidly energy is dissipated from the moving system, the more rapid the decay of oscillation and the shorter the time constant.

1.5.4   Coupled Systems

One oscillating system can act as a driving source for another one of a different frequency, as shown in figure 1-5. In this example, one system executes 18 vibrations in the time that the other system executes 16 vibrations, and the interference between them sets up a pattern of reinforcements and cancellations, which occur at a rate equal to the difference in the two frequencies, or twice per second. In auditory terms, we refer to the slow undulations as *beats*. This subject will be discussed further in chapter 2.

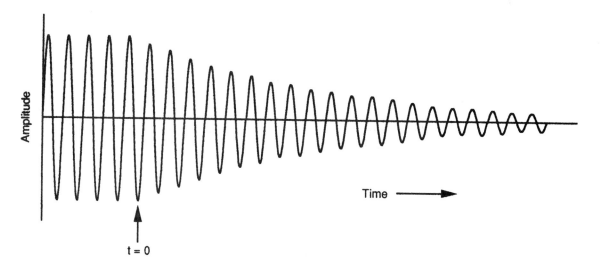

Figure 1-4. Illustration of damped oscillations.

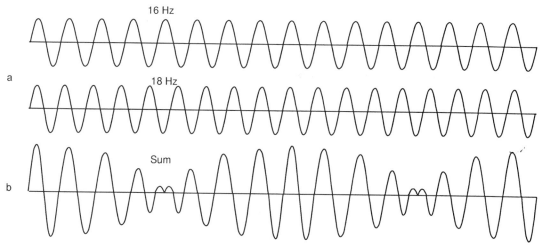

Figure 1-5. Beats between two sine waves of slightly different frequency: (*a*) 16 and 18 Hz; (*b*) their sum.

1.6   WAVE PROPAGATION

If an oscillating system is coupled to an appropriate medium, then the energy in the oscillations can be coupled to the medium and be radiated, or propagated, through it in the form of waves.

Here is a familiar example: A pebble dropped into a pool of water sets up a series of outwardly radiating rings. The water at the point of impact oscillates up and down, and its energy of oscillation is imparted to the water around it. While the waves appear to move outward, the water essentially is only moving up and down. The motion of the water is at right angles, or *transverse*, to the direction of wave propagation.

Now let us take an acoustical example. If the stem of a tuning fork is placed against a flexible, thin surface, we will hear its oscillations more clearly than if the tuning fork is simply held in the hand. The tuning fork has been coupled to the surface, and the flexible surface provides more efficient radiation of sound waves than the tuning fork alone.

The propagation of sound through air is complex, but we can provide an intuitive picture of what is going on. In figure 1-6a the air medium is shown one-dimensionally as an assemblage of small masses connected by small springs. When a single mass is displaced, as shown in figure 1-6b, the displacement is transferred to the adjacent mass through the spring, and then to the next mass, as shown in figure 1-6c.

Both compressions and rarefactions of the transmitting medium are observed during the process, and it is obvious that the motion of the small masses is in the same direction as the wave propagation, creating a *longitudinal* wave. (Recall that the wave motion in the pebble-in-water case was transverse to the water motion.)

The wave progresses through the air medium at a velocity that is dependent on the physical constants of air, as well as its pressure and temperature. In our normal experience, the dependence on pressure is fairly small, but the dependence on temperature is significant. The following equation gives the velocity of sound in air as a function of temperature:

$$\text{Velocity} = 331.4 + 0.607T \quad \text{m/sec} \tag{1.11}$$

where T is temperature in degrees Celsius.

At normally encountered temperatures, we can assume that the velocity of sound in air is 344 m/sec, which is equivalent to 1130 ft/sec.

A more general two-dimensional picture of sound propagation in air is shown in figure 1-7. Here, we have taken a snapshot, so to speak, of sound at some instant. The source oscillations have a frequency of f, and the velocity of sound in the medium is assumed to be c, in meters per second. The *wavelength*, λ, is then given by the following equation:

$$\lambda = \frac{c}{f} \quad \text{m} \tag{1.12}$$

where f is the frequency of oscillation in hertz.

The following list shows the velocity of sound (m/sec) as measured through a variety of materials.

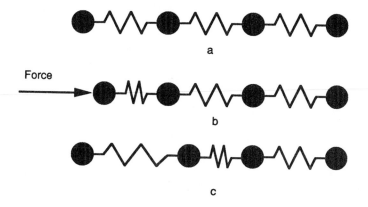

Figure 1-6. One-dimensional representation of wave motion in a medium.

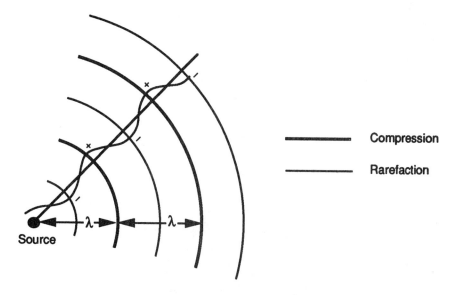

Figure 1-7. Two-dimensional representation of wave motion in a medium.

Steel	5029	Concrete	3231
Aluminum	4877	Water	1433
Brick	4176	Lead	1158
Hardwood	3962	Cork	366–518
Glass	3962	Air	344
Copper	3901	Rubber	40–149
Brass	3475		

The wavelength is the distance between successive beginnings of the waveform as the wave is propagated through a given medium.

Obviously, for larger values of f, λ will be smaller, and the following equations can be derived from equation 1.12

$$c = f\lambda \tag{1.13}$$

$$f = \frac{c}{\lambda} \tag{1.14}$$

The normal frequency range of audible sounds is from 20 Hz to 20,000 Hz (20 kHz). For comparison, note that the wavelength in air for a 20-Hz signal is 17.2 m, while that of a 20-kHz signal is 17.2 mm.

Figure 1-8 shows the frequency ranges of various instruments as related to a piano keyboard. Note the convention of using both letters and numbers to indicate each note of the keyboard. C_4, for example, denotes middle C, while C_5 denotes the octave above. C_0 is taken as the lowest note (32-foot C) on a pipe organ. A_0 is normally the lowest note on the piano. This convention will be used throughout the book.

1.7   SOUND FIELDS

Sound waves react with the acoustical environment in many ways, and the term *sound field* describes this interaction. The nature of the sound field depends on the size of the radiating source, the distance from the radiating source, obstacles in the path of sound propagation, and any patterns of sound reflection or absorption in the vicinity of the radiator.

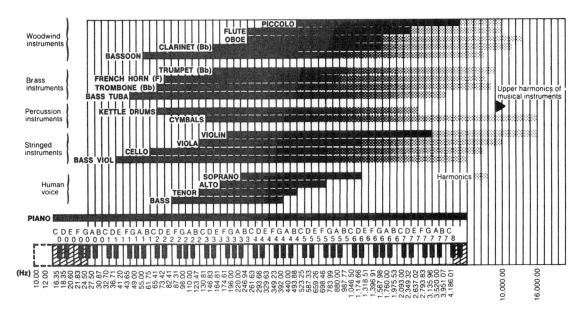

Figure 1-8. Frequency ranges of musical instruments and the human voice.

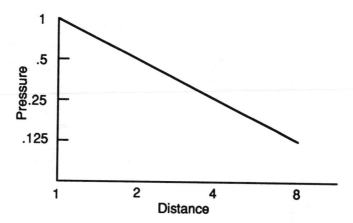

Figure 1-9. Relation between pressure and distance from a sound source in a free field.

1.7.1   The Free Field

A *free field* is an environment in which sound waves may be propagated in all directions without reflections or obstructions. The term *direct field* is often used to describe this. If we are sufficiently far away from a source, or if that source is sufficiently small, we are in the *far field* of that source. In the far field, sound pressure is observed to diminish by one-half each time we double our distance from the source. In a clear outdoor space, we can observe free-field/far-field phenomena by placing a small loudspeaker at a distance, say 3 or 4 m, above the ground, powering it with a constant signal, and measuring its sound pressure at various distances. The relation between sound pressure and distance in the free/far field is shown in figure 1-9.

Most loudspeaker and microphone manufacturers have large anechoic chambers in which to carry out free-field measurements on their products. Generally, the requirements of a free field are not easy to come by, and anechoic chambers that can provide free-field response at low frequencies are expensive. Figure 1-10 shows a sketch of an anechoic chamber. Note the deep fiberglass wedges that line all six sides of the rectangular space. The wedge depth must be on the order of a quarter wavelength or larger if the wedge is to be properly absorptive at that wavelength, so it can easily be seen that anechoic chambers that are effective down to the 50- or 100-Hz range will be quite large.

A variant of the free field is often referred to as the *open field*. One boundary of the open field is reflective, and a loudspeaker can be mounted on that surface. The remaining surfaces are absorptive, so that radiation is directed into hemispherical space without reflections or obstacles. Many such examples are constructed out of doors, as shown in figure 1-11.

1.7.2   The Near Field

If we observe a sound source at a fairly short distance in the free field, we will note some departures from the inverse pressure-distance rule described in the preceding section. What

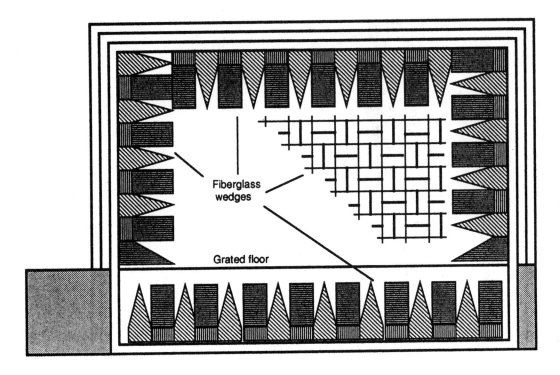

Figure 1-10. Details of an anechoic chamber. Section view.

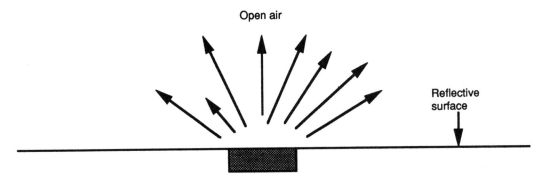

Figure 1-11. The open field.

we observe may look something like figure 1-12*a*, in which closer-in observations do not necessarily indicate higher pressures.

Very generally, the division point between the near and far fields can be determined by the information shown in figure 1-12*b*. Note that within a calculated distance from a line source of sound, the 2-to-1 pressure-distance relationship becomes a $\sqrt{2}$-to-1 relationship. In figure 1-12*c*, in the case of a large plane source of sound within a calculated distance, the pressure does not fundamentally change with observation distance.

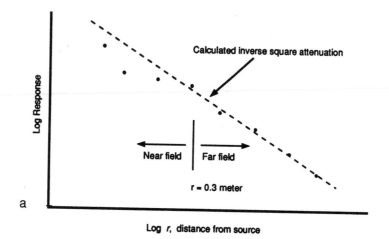

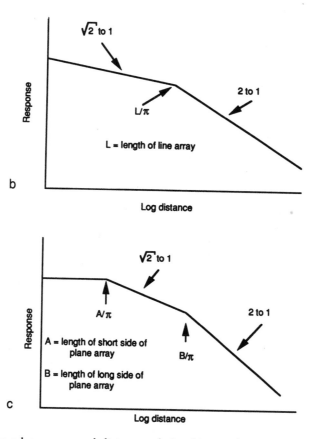

Figure 1-12. Sound pressure and distance relationships in the near field. (*a*) General observation; (*b*) attenuation with distance from a line source; (*c*) attenuation with distance from a plane source.

As a general rule, we can assume that we are effectively in the free field of a device beyond the distance A/π meters, where A is the largest dimension in meters across the front of the radiating surface. Thus, for a loudspeaker 2 m tall, we would be effectively in the far field at a distance of about 0.64 m.

1.7.3 Obstacles in the Path of Sound Radiation: Diffraction, Reflection, and Shadowing

When we place an obstacle in the path of sound radiation in a free field, what happens is dependent on the wavelength of the sound. At figure 1-13a, an object that is small relative to the radiated wavelength is virtually "invisible" to the propagation of sound. Sound waves are said to *diffract*, or bend, around the obstacle and continue their propagation as before.

If we keep the obstacle the same size, but shorten the wavelength, as shown in figure 1-3b, we get into a more complex situation. Some of the sound is reradiated by reflection from the obstacle, while some is diffracted around it.

In figure 1-13c, at a still shorter wavelength, most of the sound striking the obstacle is reflected, and a shadow zone is created behind it.

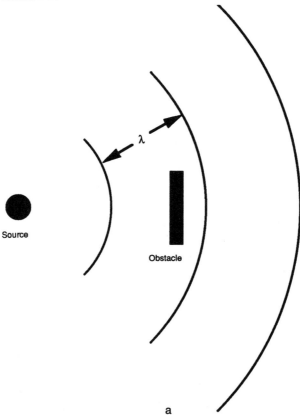

Figure 1-13. Obstacles in the path of sound waves: (*a*) at long wavelengths; (*b*) at medium wavelengths; (*c*) at short wavelengths.

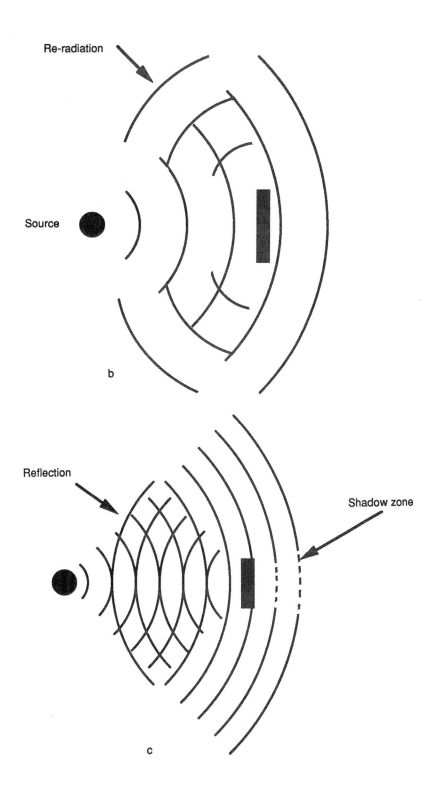

Re-radiation

Source

b

Reflection

Shadow zone

c

15

All these observations are consistent with everyday experience. If we listen to someone talk face to face, we hear all the voice frequencies in correct proportion. If that person turns around and continues to talk, we will hear the low frequencies of speech as before, since they diffract around the obstacle, which in this case is the talker's head. At the highest frequencies, the shadowing effect of the head will be significant, and we will hear those frequency components less clearly.

The nature of reflective surfaces is of great importance in architectural acoustics, and figure 1-14 gives some indication of this. In figure 1-14a we see what is called specular reflection. Short wavelengths are reflected very much the way light is from a mirror, with the angle of reflection essentially the same as the angle of incidence. In figure 1-14b we see an example of staggered reflections from equally spaced planes. The effect is often an unpleasant one,

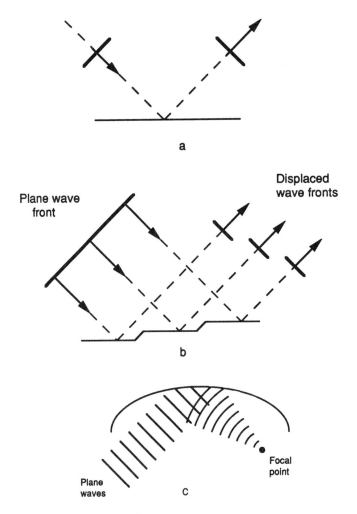

Figure 1-14. Reflective surfaces. (a) Specular reflection from a flat surface; (b) staggered reflections from equally spaced planes; (c) focused reflections from a concave surface; (d) diffuse reflections from a quadratic residue diffuser.

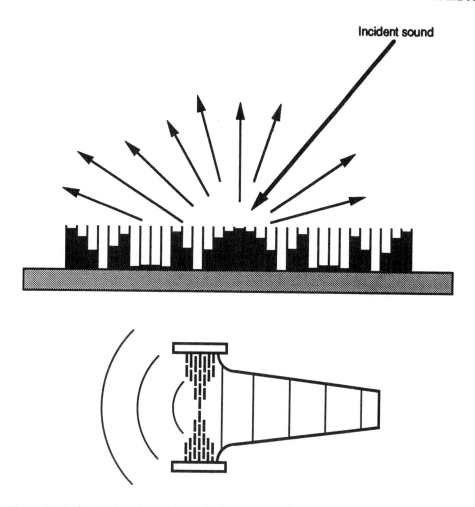

Figure 1-15. Illustration of a perforated-plate acoustic lens.

reinforcing some frequencies and canceling others (see section 1.11). A large concave spherical surface can focus plane waves into a pronounced "hot spot," as shown in figure 1-14c.

While we can intuitively see how sound is reflected from a flat or concave surface, the action of the *quadratic residue diffusor*, shown in figure 1-14d, is not easy to see. The wells in the diffusor, depending on their depth and spacing, interact with the incident sound waves, producing reflections in one plane that are primarily independent of the angle of incidence over a segment of the frequency range.

Sound diffracts, or bends around, obstacles that are suitably small relative to wavelength. The acoustic lens, as used in some loudspeaker systems, is an example of this. The perforated-plate lens shown in figure 1-15 creates a shorter path length for sounds passing through the middle of the structure. Thus, plane waves diverge outward as they pass through the lens plates.

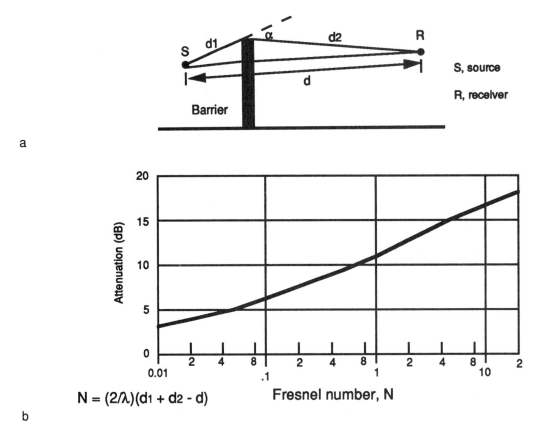

Figure 1-16. Sound diffraction over the top of an extended barrier. (a) Physical view (S is source and R is receiver); (b) graph for reading attenuation.

Another example is shown in figure 1-16a. Here, a tall structure, such as a barrier erected by a noisy highway, will attenuate high-frequency sounds by providing an especially steep angle for sound to bend around. The amount of attenuation can be quite accurately estimated for an incoherent line source, such as a highway, as indicated in figure 1-16b.

1.7.4 Absorptive Barriers in the Sound Field

Figure 1-17 shows the effect of an absorptive barrier in the sound field. At long wavelengths, as shown in figure 1-17a, some of the sound is absorbed and transmitted on by the barrier, while the remaining sound is reflected. In figure 1-17b, we illustrate the wavelength-dependent nature of absorption. Here, we show the effect of waves incident on and reflecting from a hard surface. Note that at a distance of one-fourth wavelength from the surface, the air particle velocity is high. If we wish to absorb sound effectively at the wavelength in question, then the absorptive material must have a thickness that is a substantial portion of one-quarter wavelength.

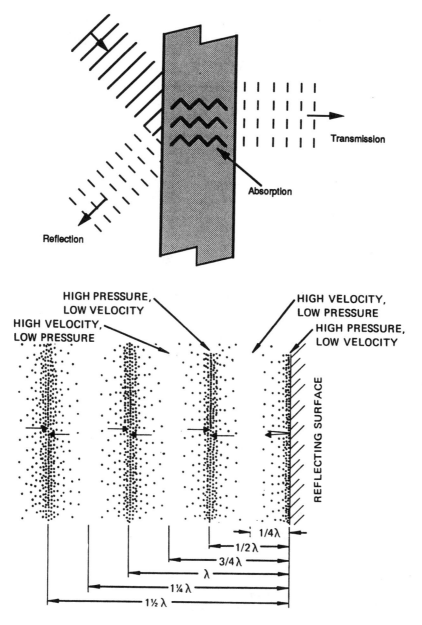

Figure 1-17. Absorptive barriers. (*a*) Absorption, transmission, and reflection of sound; (*b*) wavelength relationships at a reflective surface. A plane wave reflected from a plane surface at normal incidence produces well-defined zones of high pressure alternating with zones of high particle velocity at distances of one-quarter wavelength. Data *a* and *b* courtesy of JBL Inc.

In the above cases, we can measure a sound *absorption coefficient*, which determines how much sound is absorbed by the barrier and how much is reflected. The Greek letter alpha (α) is used to indicate the ratio of sound absorbed by the barrier.

Over the years, detailed measurements have been made of sound absorption coefficients, as a function of frequency, for a wide variety of materials. Many materials have absorption coefficients approaching 0.95, indicating that only 5% of the incident sound is reflected. Other materials—concrete walls, for example—may have an absorption coefficient of 0.05, indicating that 95% percent of the incident sound will be reflected. Most values range between these extremes.

1.7.5   The Diffuse Field

A *diffuse* field is one in which sound pressure is the same in all locations and in which sound propagation is as likely in one direction as in any other. The simplest diffuse field is the reverberant field, which exists when there are many reflecting surfaces surrounding a sound radiator. A typical case may be observed indoors in a space that has predominantly reflective boundaries (with an average absorption coefficient of about 0.2 or less) and essentially irregular, nonparallel surfaces. When these conditions hold, a sound source, when it is turned on, will exhibit a sound pressure growth curve as shown in figure 1-18. When the steady-state condition is reached, reflected sound in the room is as likely to be incident from one direction as from any other. During the steady-state portion, sound energy is absorbed or dissipated at the room boundaries at the same rate as it is generated by the sound source.

When the source of sound is turned off, the decay of the reverberant field is caused by the gradual dissipation of the remaining acoustical energy in the room.

When we speak of "liveness" in a room, we are referring to *reverberation* and the diffuse field it creates in the space. In describing reverberation, we customarily define what is called *reverberation time*, the time measured in seconds for the sound pressure in the diffuse field to diminish to one-thousandth its steady-state value after the sound source has been turned off.

The Sabine equation for reverberation time is given below:

$$T = \frac{0.16V}{S\overline{\alpha}} \tag{1.15}$$

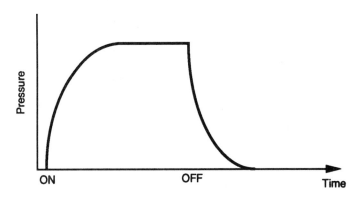

Figure 1-18. Growth and decay of sound in a reverberant space.

where V is the volume of the room, measured in cubic meters, S is the surface area of all room boundaries, measured in square meters, and $\bar{\alpha}$ is the average absorption coefficient of all internal room boundaries.

A room must be quite large and acoustically live in order to exhibit a diffuse reverberant field at low frequencies. The problem is, of course, that low frequencies have long wavelengths, and long wavelengths require space in which to develop and propagate. Schroeder developed the following equation, which essentially gives the frequency above which the diffuse field exists:

$$f = 2000 \sqrt{\frac{T}{V}} \qquad (1.16)$$

where T is the reverberation time in the space and V is the volume in cubic meters.

1.7.6 Relation between the Direct and Reverberant Fields

If we place a sound source, such as a loudspeaker, in a fairly live room, any observations at distances close to the loudspeaker will be dominated by the direct field produced by the loudspeaker. As we move away from the loudspeaker, the direct field will diminish in sound pressure by a factor of one-half with each doubling of distance. Eventually, we will come to a point at which the direct field and the diffuse field are more or less equal; that distance from the loudspeaker is known variously as the *reverberation radius* or the *critical distance*. Beyond that point, it is clear that the reverberant field will be dominant. This important relationship is shown in figure 1-19.

For an omnidirectional radiator, the critical distance D_C is given by

$$D_C = 0.14 \sqrt{S\bar{\alpha}} \qquad (1.17)$$

where S is the surface area in the space and $\bar{\alpha}$ is the average absorption coefficient. The equation is consistent in either metric or English units.

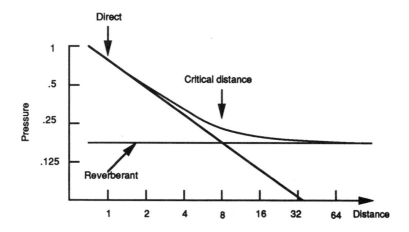

Figure 1-19. Relation between direct and reverberant fields as a function of source distance.

In later chapters dealing with performance spaces, we will expand our discussion of sound fields extensively.

1.8   POWER, SOUND PRESSURE, AND THE DECIBEL

In any aspect of sound or acoustical technology, the *decibel* is universally used to measure acoustical power and pressure levels. Fundamentally, the *bel* is defined as the logarithm (base 10) of a simple power ratio:

$$Bel = \log \left(\frac{W_1}{W_0} \right) \tag{1.18}$$

In this equation, W_0 is a reference power, and W_1 is any other power whose level, relative to the reference power, we wish to solve for.

The decibel is defined as

$$Decibel = 10 \log \left(\frac{W_1}{W_0} \right) \tag{1.19}$$

Reference power W_0 may be taken as any convenient value, but in general audio engineering, there are two references used, one watt and one milliwatt. Levels calculated relative to one watt are expressed in dBW, while levels calculated relative to one milliwatt are expressed in dBm.

The term *level* is generally used to indicate values expressed in decibels. The virtue of dealing with levels is that we use a relatively small range of numbers to describe a very large range of actual powers and sound pressures. For example, figure 1-20 shows typical powers and power levels calculated relative to one watt. Note that a range of 60 dB corresponds to a power ratio of 1,000,000 to 1.

Since sound pressure is such a useful quantity, and relatively easy to measure, we can adapt the decibel accordingly. Fundamentally, sound pressure level (SPL or L_p) is defined as

$$L_p = 20 \log \left(\frac{P_1}{P_0} \right) \tag{1.20}$$

where P_0 is established as 2×10^{-5} Pa, which is quite close to the normal threshold of hearing. The introduction of the multiplier of 20 in this equation comes about because power is proportional to the *square* of pressure. Figure 1-21 shows representative sound pressures and their corresponding levels.

When we add levels, we do not simply add their numerical values. We must, in effect, convert back to power or pressure, add those values, and then convert to level once again. The nomograph shown in figure 1-22 allows us to do this by inspection. Note that if two levels differ by more than about 10 dB, then the sum of the two is effectively the same as the higher of the two values.

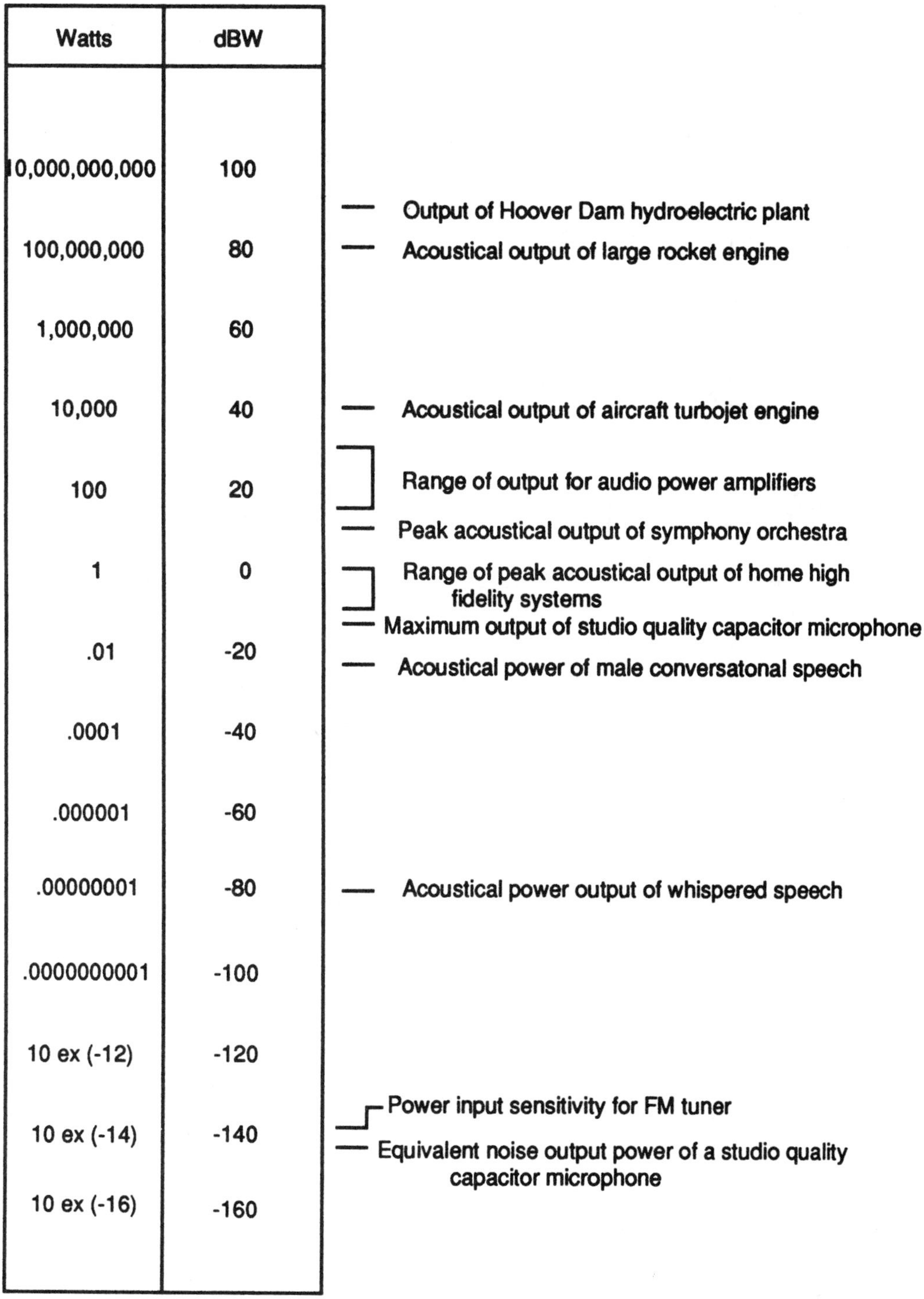

Watts	dBW
10,000,000,000	100
100,000,000	80
1,000,000	60
10,000	40
100	20
1	0
.01	-20
.0001	-40
.000001	-60
.00000001	-80
.0000000001	-100
10 ex (-12)	-120
10 ex (-14)	-140
10 ex (-16)	-160

— Output of Hoover Dam hydroelectric plant
— Acoustical output of large rocket engine

— Acoustical output of aircraft turbojet engine

⌐ Range of output for audio power amplifiers

— Peak acoustical output of symphony orchestra

⌐ Range of peak acoustical output of home high fidelity systems
— Maximum output of studio quality capacitor microphone
— Acoustical power of male conversatonal speech

— Acoustical power output of whispered speech

⌐ Power input sensitivity for FM tuner
— Equivalent noise output power of a studio quality capacitor microphone

Figure 1-20. Representative acoustical powers and power levels.

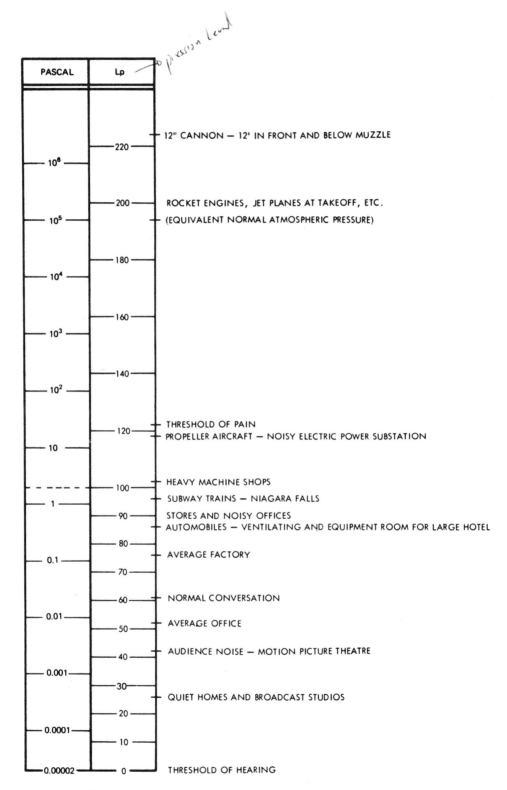

Figure 1-21. Representative acoustical pressures and pressure levels.

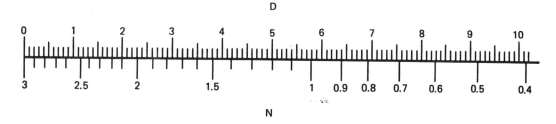

Nomograph for Adding Levels Expressed in dB. Summing Sound Level
Output of Two Sound Sources Where **D** is Their Output Difference in dB,
N, is Added to the Higher Level to Derive Total Level

Figure 1-22. Summing levels in decibels. D = difference between two levels; N = quantity to be added to the greater of the two levels.

1.9   DIRECTIONAL CHARACTERISTICS OF SOUND RADIATORS

Musical sound sources generally have very complex directional radiating characteristics, but fundamentally these characteristics are dominated by wavelength considerations. Low frequencies tend to be more or less omnidirectional, since the sound sources that produce them are small relative to a wavelength. At higher frequencies (and shorter wavelengths), preferences for directional radiation show up in many ways.

For the engineer, quantifying the directional charactistics of a loudspeaker, microphone, or musical instrument is important, and in acoustical engineering, the concept of *directivity index* (DI) has been developed. It is shown graphically in figure 1-23. Here, an omnidirectional source of known acoustical power output is used as a reference. L_p is noted at a given distance. Then, a directional radiator whose acoustical conversion efficiency has been carefully measured produces the same acoustical output power, and the new value of L_p is measured along the major axis of the device. The difference in the two readings of L_p is the directivity index of the device along its major radiating axis. A device that radiates evenly in all directions has a DI of 0 dB, while an extremely directional radiator, such as a loudspeaker intended for sound reinforcement "long throw" applications, may have a DI value in the range of 15 to 18 dB.

A related quantity is *directivity factor* (DF or Q)

$$DF = 10^{DI/10} \tag{1.21}$$

Figure 1-24 shows the wavelength dependence of directionality for sound emanating from the end of a long tube. The condition is roughly analogous to the bell of a trumpet or other brass instrument. The diameter of the tube, for each frequency, is expressed as a portion of the radiated wavelength. The circular grid lines are spaced at intervals of 10 dB.

It is apparent that when the tube diameter is 0.16 wavelength, then radiation from the tube is virtually omnidirectional. Moving on to the fifth example in the set, when the diameter of the tube is equal to a wavelength, the DI has reached a value of 9.6 dB, and the radiation is more or less confined to a cone with a beamwidth of about 90°. (In acoustical engineering, beamwidth

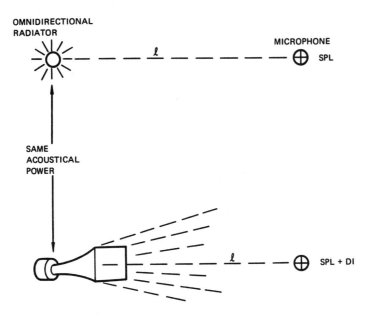

OMNIDIRECTIONAL
RADIATOR

MICROPHONE

ℓ

SPL

SAME
ACOUSTICAL
POWER

ℓ

SPL + DI

Figure 1-23. Illustration of directivity index (DI).

is generally defined as the included angle over which radiation is no less than -6 dB relative to the on-axis reference of 0 dB.)

Let us consider the case of a trumpet. The bell is roughly 0.1 m in diameter, and this corresponds to one wavelength at a frequency of 3440 Hz. Thus, at that frequency, the trumpet would have a DI of 9.6 dB. But, noting the data in figure 1-8, that frequency is well outside the range of fundamental tones that can be played on the trumpet. According to that figure, a fundamental frequency of 1 kHz is about as high as the instrument can be played, and the corresponding directional plot describing that would be when the tube diameter was equal to 0.32 wavelength. And at that frequency the instrument is virtually omnidirectional.

What these data show us is that the fundamental tones produced by the trumpet are all virtually omnidirectional. Only the higher harmonics and overtones of the instrument have a strong bearing on its directional character. This nature of the instrument will be discussed in chapter 7.

Related data are shown in figure 1-25 for a piston mounted in a large wall. Here, the wavelength range is 24 to 1. At the highest frequencies, we can see that there are minor lobes in the polar response at the edges of the useful radiation angle.

An unbaffled piston acts as a *dipole*, radiating identical patterns at both zero and 180°, with no output at 90° and 270°. It is important to note that there is always a 180° phase relationship between front and back radiation with a dipole: when one side is "pushing," the other side is "pulling." We will discuss this further in our analysis of directional microphones in a later chapter. Some percussion instruments, the tambourine in particular, approximate dipole action. Typical directional patterns are shown in figure 1-26.

Our discussion of directionality will be considerably expanded as we study the various musical instruments in later chapters.

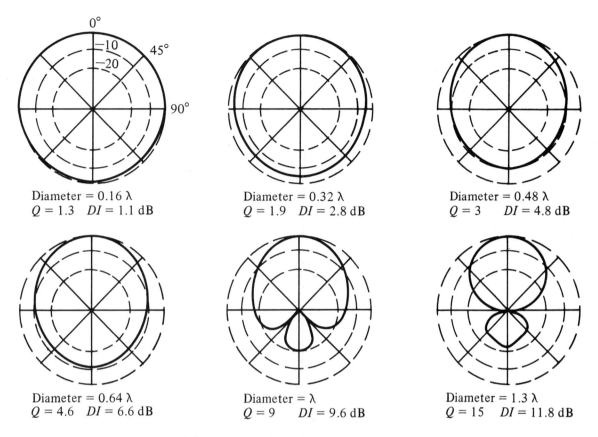

Figure 1-24. Directional characteristics for a piston in the end of a long tube.

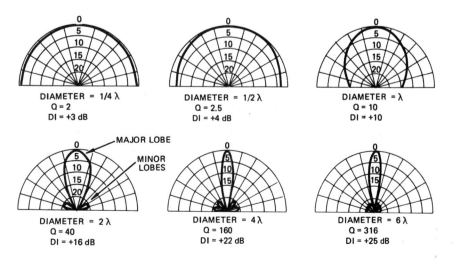

Figure 1-25. Directional characteristics for a piston in a large wall.

1.10   ENVIRONMENTAL ACOUSTICAL EFFECTS

There are many environmental effects that can profoundly influence the way we hear music out of doors, and in some cases indoors. Some of the more important ones are discussed here.

1.10.1   Effect of Temperature Gradients

The effect shown in figure 1-27*a* may be noticed shortly after sundown, when the ground is still warm but the air above is cool. Since the velocity of sound is greater in warm air than in cool, there will be a shifting of direction of propagation as shown. The effect shown in figure 1-27*b* can be noticed early in the morning, when the ground is still cool but the air above has been warmed.

1.10.2   Effects of Wind Velocity Gradients

Wind velocity is usually greater several meters above the ground than it is at ear level, and the effect shown in figure 1-28*a* is typical. The effect is different, depending on whether the listener is upwind or downwind. In strong breezes, there can be pronounced skipping and dead zones.

   The case shown in figure 1-28*b* is typical of a strong cross breeze. The velocity of sound is the vector sum of the velocity in still air plus the velocity of the air itself. When the air velocity is greatest from the side, then the apparent directionality of large radiators—loudspeakers at an outdoor concert, for example—can be shifted accordingly.

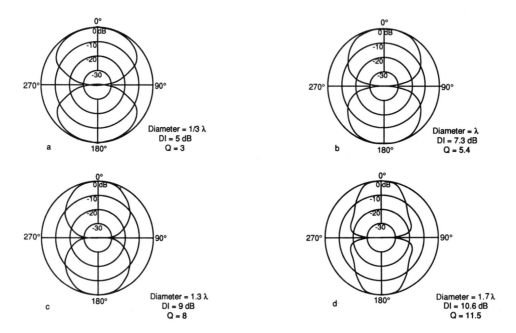

Figure 1-26. Directional characteristics for an unbaffled piston (dipole).

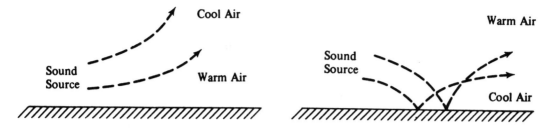

Figure 1-27. Effect of temperature gradients on sound transmission.

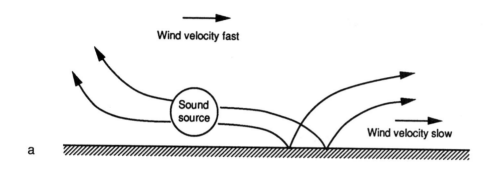

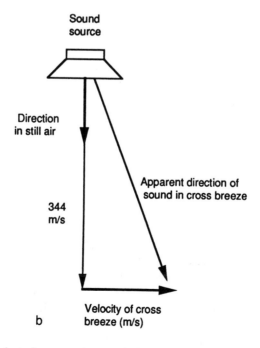

Figure 1-28. Effect of wind on sound transmission.

1.10.3   Doppler Frequency Shift

We have all heard the familiar sound of an automobile horn rising in pitch as the vehicle approaches us, followed by a falling in pitch as it moves away from us, both relative to the pitch of the horn when the vehicle is at rest.

The basic phenomenon is as shown in figure 1-29. In front of and in line with the moving source, the wavelengths have been crowded together, while behind the source they have been spread apart. The shorter wavelengths correspond to a higher frequency, and the longer ones to a lower frequency.

As a simple example, let us assume that a stationary sound source has a wavelength of 1 m; that is, it is emitting a frequency of 344 Hz, and the velocity of sound is of course 344 m/sec. Let the source of sound move toward the listener at a velocity of 34.4 m/sec. Then, the wavelength in front of the source will be

$$\lambda = \frac{344 - 34.4}{344} \text{ m}$$

$$\lambda = 0.9 \text{ m}$$

The wavelength of 0.9 m corresponds to the frequency

$$f = \frac{344}{0.9} = 382 \text{ Hz}$$

Behind the moving sound source, the wavelength will be

$$\lambda = \frac{344 + 34.4}{344} \text{ m}$$

$$\lambda = 1.1 \text{ m}$$

The wavelength of 1.1 m corresponds to a frequency of

$$f = \frac{344}{1.1} = 313 \text{ Hz}$$

1.10.4   Effects of Humidity

Most people intuitively believe that music sounds less lively and bright when the weather is humid. In fact, the opposite is true. What is dominant here is the way people generally feel under such conditions. There are substantial atmospheric losses at high frequencies, and these are most apparent in the range of 15 to 20% relative humidity. The data shown in figure 1-30 allow us to calculate these losses. For example, an 8-kHz tone at 50% relative humidity will undergo an air loss of 3 dB for each 30 m of travel. These losses are in addition to the normal inverse square loss of sound pressure traveling over distance. Thus, music may sound

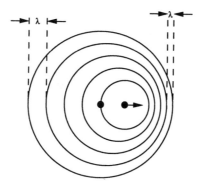

Figure 1-29. Doppler frequency shift.

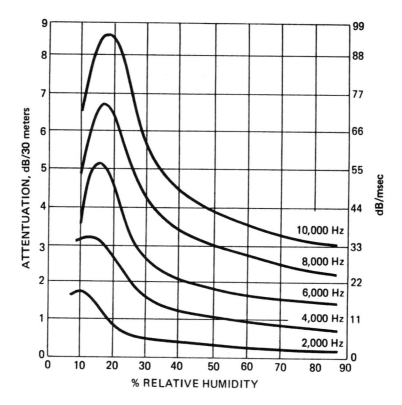

Figure 1-30. Sound absorption in air as a function of frequency and relative humidity.

dull if listened to over large distances, especially when the relative humidity is in the range of 20%.

Figure 1-31 gives another view of the same phenomenon. Here, we see the effect of overall attenuation resulting from both inverse square losses and high-frequency humidity losses.

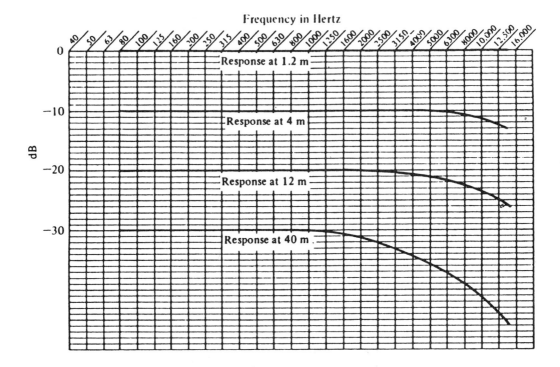

Figure 1-31. Losses due to both inverse square (distance) relationships and high-frequency attenuation (at 20% relative humidity and 20° Celsius).

1.11   INTERFERENCE EFFECTS

Interference effects occur when sound sources are combined through some kind of delay path. The delay can come about in several ways. Discrete reflections, or echoes, are quite common. And in some cases involving microphones in recording or sound reinforcement, two or more microphones at different distances may pick up the same sound; when these are combined into a single channel, the result is signal cancellation and reinforcement.

A common example is shown in figure 1-32a. Here, two loudspeakers have been placed a distance of 6 m apart. Listeners are located at A (0.6 m from the center line) and at B (3 m from the center line). The slight lateral displacement of position A produces the pattern of signal reinforcements and cancellations shown by the dashed line in the graph.

The subjective effect is usually worse for small values of off-center listening than for larger ones, inasmuch as the interference notches in response are quite wide, representing large losses in signal.

Figure 1-32b shows the actual response (solid line) and the averaged one-third octave response (dashed line) for the listener located at position B. Here, the one-third-octave averaging fills in many of the dips in response, and this is more the way we would hear them, based on critical band theory, a subject to be discussed in the next chapter.

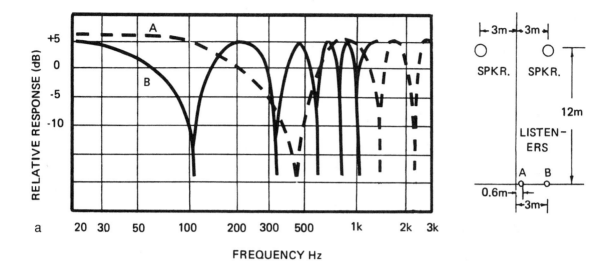

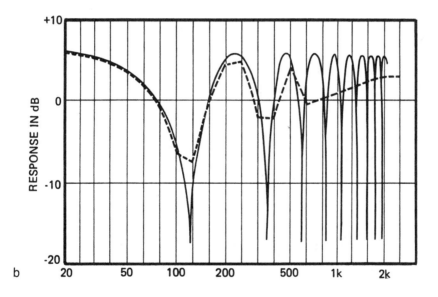

Figure 1-32. Interference effects from two sound sources observed at two positions. Solid line = measured sine-wave frequency response; dotted line = one-third octave band response, closely corresponding to subjective tonal quality when listening to normal program material. Above 1 kHz, subjective response is essentially flat. Data courtesy of JBL Inc.

REFERENCES

Backus, J. 1969. *The Acoustical Foundations of Music*. New York: W. W. Norton.

Benade, A. 1976. *Fundamentals of Musical Acoustics*. London: Oxford University Press.

Beranek, L. 1954. *Acoustics*. New York: McGraw-Hill.

————. 1962. *Music, Acoustics and Architecture*. New York: Wiley.

Berg, R., and D. Stork. 1982. *The Physics of Sound*. Englewood Cliffs, NJ: Prentice-Hall.

Campbell, M., and C. Greated. 1987. *The Musician's Guide to Acoustics*. New York: Schirmer Books.

Cremer, L., and H. Mueller. 1982. *Principles and Applications of Room Acoustics*. Translated by T. Schultz. London: Applied Science Publishers.

Doelle, L. 1972. *Environmental Acoustics*. New York: McGraw-Hill.

Eargle, J. 1986. *Handbook of Recording Engineering*. New York: Van Nostrand Reinhold.

Kinsler, L., et al. 1982. *Fundamentals of Acoustics*. New York: Wiley.

Knudsen, V., and C. Harris. 1978. *Acoustical Designing in Architecture*. New York: Acoustical Society of America.

Kuttruff, H. 1979. *Room Acoustics*. London: Applied Science Publishers.

Moravcsik, M. 1987. *Musical Sound*. New York: Paragon House Publishers.

Morse, P. 1948. *Vibration and Sound*. New York: McGraw-Hill.

Olson, H. 1952. *Musical Acoustics*. New York: McGraw-Hill.

Pierce, J. 1983. *The Science of Musical Sound*. New York: W. H. Freeman.

Rathe, E. 1969. "Notes on Two Common Problems in Sound Propagation," *J. Sound & Vibration* 10:472–479.

Rossing, T. 1990. *The Science of Sound*. Reading, MA: Addison-Wesley.

Schroeder, M. 1984. "Progress in Architectural Acoustics and Artificial Reverberation: Concert Hall Acoustics and Number Theory." *J. Audio Engineering Society* 32(4):194–203.

Strong, W., and G. Plitnik. 1979. *Music, Speech & High Fidelity*. Provo, UT: Brigham Young University Press.

2

Psychological Acoustics

The field of psychological acoustics, or psychoacoustics, is extremely broad, encompassing many disciplines. Our approach here is to limit the subject to aspects of special interest to those involved in music technologies, such as recording, broadcasting, sound reinforcement, and electronic music.

In this chapter we will discuss some aspects of hearing having to do with our response to certain test stimuli. We will begin with a brief review of the physiology of the ear, then move on to a discussion of loudness phenomena. Localization, both in the natural binaural environment and via loudspeakers and headphones, will be discussed, as will signal processing techniques for image broadening and fusion. Pitch perception will be dealt with, and consonance and dissonance will be discussed in terms of critical band theory.

2.1 PHYSIOLOGY OF THE EAR

Figure 2-1*a* shows a physical view of the ear. Note that there are three regions: the outer, middle, and inner ear. The outer visible portion of the ear, called the *pinna*, leads through the *meatus* to the *tympanic membrane*, or *eardrum*, a distance of about 27 mm (1 in). The pinna plays a very important function in localization at high frequencies. Specific resonances and reflections in the convolutions of the pinna help us determine the direction of high-frequency sounds in both fore-aft and height directions.

The middle ear begins at the tympanic membrane. Three tiny bones, the *malleus*, *incus*, and *stapes* (known as the hammer, anvil, and stirrup), transmit the motion of the eardrum to the inner ear. The arrangement of bones acts as a mechanical transformer, providing an increase in sound pressure at the entrance to the inner ear relative to that at the eardrum. The *eustachian tube* connects the middle ear with the throat to equalize long-term pressure differentials between the atmosphere and the inner ear. Through added muscular control, the bones of the middle ear can limit some of the effects of very loud sounds, thus protecting the inner ear from possible damage.

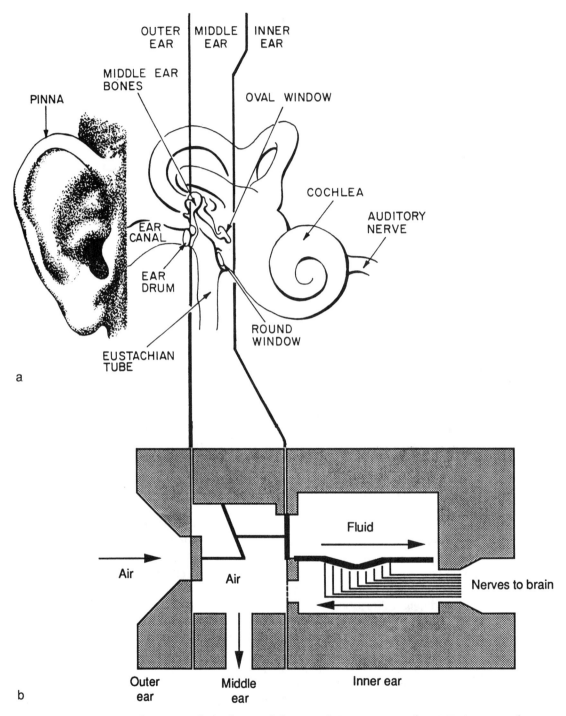

OUTER EAR | MIDDLE EAR | INNER EAR

PINNA

MIDDLE EAR BONES

OVAL WINDOW

COCHLEA

AUDITORY NERVE

EAR CANAL

EAR DRUM

ROUND WINDOW

EUSTACHIAN TUBE

a

Fluid

Air

Air

Nerves to brain

Outer ear

Middle ear

Inner ear

b

Figure 2-1. Structure of the ear. (a) Physical view; (b) functional view. Data at a after C. T. Morgan et al., *Human Engineering Guide to Equipment Design*, courtesy McGraw-Hill, New York.

The inner ear begins at the oval window, which is directly acted upon by the stapes. The oval window is at one end of the *cochlea*, which is the central organ of the inner ear. As is seen in figure 2-1*a*, the cochlea resembles a snail's shell. In the functional view shown in figure 2-1*b*, it has been straightened so that its structure can be seen more clearly. The *basilar membrane* runs the length of the cochlea, some 35 mm (1.4 in.). The cochlea is filled with fluid on both sides of the basilar membrane, and there is a small opening, called the *helicotrema*, at the far end to equalize fluid pressure on both sides of the membrane. There is a region along the basilar membrane, known as the organ of Corti, which contains about 30,000 nerve endings equally spaced along its length. These nerve endings are part of the auditory nervous system, and the information from both ears is communicated directly to the brain.

It has been determined that certain locations along the basilar membrane vibrate predominantly in sympathy with certain frequencies; this is the so-called *place theory* of audition. In this sense, the basilar membrane functions as a wave analyzer, but this is an oversimplification, inasmuch as the ear can respond as well to very short, impulsive sounds.

The resolution of the cochlea is fairly broad, corresponding to what is known as a critical band of frequencies. With pure tones this is indeed the case, and we cannot clearly separate two simultaneous pure tones that lie within a single critical band. In a musical sense, the mechanism is more acute than this, and it is possible to determine the pitch of sequential pure tones with perhaps ten times the acuity as in the case of simultaneous excitation. With complex musical waveforms, there is far more information available, and the ear-brain can sort out very small differences in frequency.

The information from both left and right ears is compared in the brain, and fine judgments regarding interaural time, frequency, and amplitude relationships provide our remarkable abilities to determine precisely the lateral location of sound sources.

2.2   LOUDNESS PHENOMENA

Figure 2-2 shows the Robinson-Dadson equal loudness contours for persons with normal hearing. These curves, which represent refinements of earlier work done by Fletcher and Munson, were determined by comparing the subjective loudness of pure tones of varying frequencies with the loudness of a 1-kHz tone. Tests were carried out at various reference levels, and the loudness contours, known as *phons*, match the actual sound pressure levels at 1 kHz.

Note that in the region of maximum acuity, the ears can detect a level range of about 120 dB, corresponding to a pressure ratio of 1,000,000 to 1. The lowest curve indicates the minimum audible field (MAF), while the highest curve indicates the threshold of feeling. At that threshold, listeners experience a tingling effect in the ears, and this is indicative of potential damage to the inner ear.

The curves show that the ear is most sensitive to sound in the 3- to 3.5-kHz region and that there is a substantial falloff in hearing sensitivity at low frequencies at low levels. Many stereophonic preamplifiers or receivers have "loudness" controls. These switch into the circuit a bass boost that is roughly the inverse of the loudness contours in the 40- or 50-phon range, thus providing a more pleasant balance between high and low frequencies for the listener who wants to play back recordings at moderate or low levels.

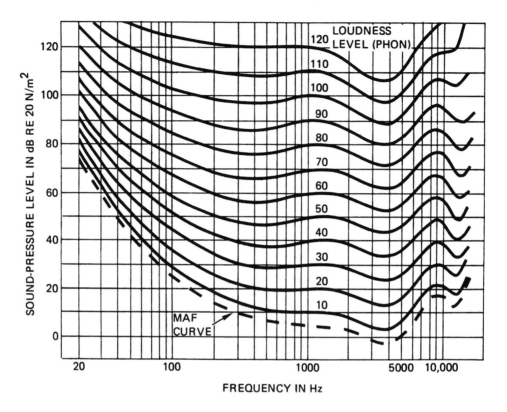

Figure 2-2. Robinson-Dadson equal loudness contours.

Loudness implications carry over into the economics of sound system design. In the typical motion picture theater, reference levels of $L_p = 85$ dB are common. If low-frequency special effects at 40 Hz are desired, then they will have to be played back at levels of 100 dB in order to be of equal loudness with signals in the 1-kHz range.

2.2.1   Measurement of Sound Pressure Level

The sound level meter (SLM) is normally used for measuring sound pressure levels. It is rigorously standardized throughout the industry, and an important aspect of these instruments is the set of weighting curves that is built into them. Figure 2-3a shows a typical SLM, with the three weighting curves shown in figure 2-3b.

The C curve is more or less flat and would be used for measuring the loudness level in phons for high-level signals. The B curve, often called the 70-phon curve, may be used for measurements in the 60- to 80-phon range. The A curve is used to measure low-level signals, in particular the low-level noise performance of recording systems and microphones. For these purposes the A curve is ideal.

The designations dB(A), dB(B), and dB(C) are often used to indicate which weighting curve has been used.

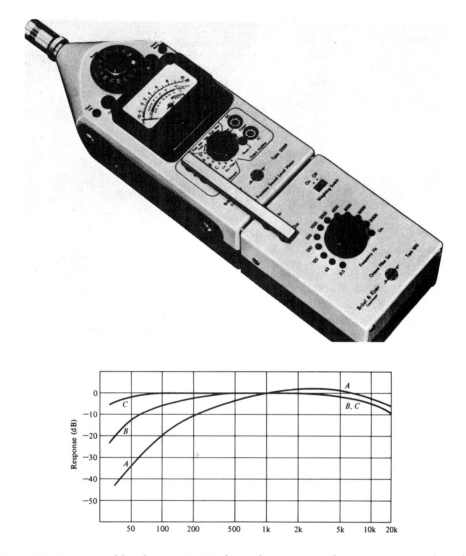

Figure 2-3. (*a*) A sound level meter (SLM); (*b*) weighting curves. Photo at *a* courtesy of Brüel & Kjaer.

2.2.2   OSHA Safety Criteria

Present-day activities in music recording and reinforcement can subject operating personnel to extremely high sound pressure levels. As we shall see later, even players of traditional musical instruments in symphony orchestras are exposed to their share of abuse. The greatest potential for damage to hearing results from prolonged exposure to noise in industrial situations, and awareness of noise in the workplace has resulted in legislation aimed at protecting citizens. The Occupational Safety and Health Act (OSHA) specifies maximum allowable daily exposure to sound in the workplace, as given in table 2-1.

Table 2-1. Permissible Exposure to Noise
Levels under OSHA Regulations

Sound Pressure Level, dB (A weighted)	Daily Exposure, h
90	8
92	6
95	4
97	3
100	2
102	1.5
105	1
110	0.5
115	0.25

Source: Occupational Safety and Health Act, 1970.

The Environmental Protection Agency (EPA) proposes even more stringent regulations, suggesting a limit of 85 dB for 8 hours of exposure daily, and a trading value of 3 dB for each doubling or halving of exposure.

2.2.3   Loudness and Loudness Level

Loudness level is measured in *phons*, a decibel scale appropriately weighted across the frequency band. Loudness itself is defined otherwise, and the unit is the *sone*. In some areas of noise assessment, loudness, as measured in sones, is more useful than loudness level. Figure 2-4 shows the relationship between loudness and loudness level. For complex signals, the actual loudness varies slightly from the data in the figure, but we will not discuss this further. The reader who wants more information in this area is referred to the many noise control handbooks that are available.

Most listeners will judge a signal that is 10 dB higher than another to be "twice as loud," and one that is 10 dB lower to be "half as loud." This is carried over into the relationship between phons and sones, and we note that each 10-dB increase in phons corresponds to a doubling of loudness in sones.

2.2.4   Loudness Versus Signal Duration

Signals of short duration do not sound as loud as those of longer duration, and figure 2-5a shows the relationship. A tone with a duration of 2 msec will sound about 15 dB lower in loudness than the same tone with a duration of 200 msec. Note that in the range of 0.2 sec and longer, tone duration has little effect on loudness level.

In a practical sense, many program metering methods used in broadcast and recording take advantage of the ear's relative insensitivity to high-level signals of very short duration. The normal sluggish ballistics of the venerable VU (volume unit) meter have worked in favor of

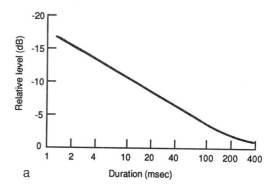

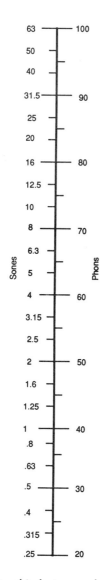

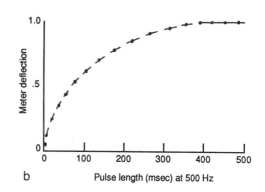

Figure 2-4. Relationship between phons and sones.

Figure 2-5. (*a*) Dependence of loudness on tone duration; (*b*) ballistics of a standard Volume Unit (VU) meter.

maintaining equal loudness in transmission channels for years. Figure 2-5*b* shows the response of a VU meter relative to that of a so-called peak program meter. Note that the ballistics of the VU meter are roughly the inverse of the data shown in figure 2-5*a*.

There is, however, a need for both loudness-related metering and absolute peak metering. The loudness approach is important in establishing consistent operating levels in broadcasting and recording, while peak analysis of the signal is important in terms of absolute modulation limits of the medium. The ear can duplicate the function of the former, but not the latter.

2.3   LOCALIZATION PHENOMENA

In general, it is difficult to localize pure tones. Those sounds that are easiest to localize in the lateral plane are complex signals, such as music and speech. The presence of short-term, or transient, components in the program makes the task easier.

The head itself provides significant shadowing of high frequencies above about 2 kHz, and this provides higher pressure at the ear closest to a laterally displaced sound source.

At frequencies below about 700 Hz, sound tends to diffract around the head, and the pressures at the ears will not be significantly different. In this frequency region and below, phase relationships at the two ears are significant in making lateral localization judgments. An excellent demonstration of this is presented in the recording listed in the Additional Resources section at the end of the chapter.

2.3.1   Binaural Hearing and Headphone Listening

Normal hearing through the two ears is termed *binaural*, and the effect can be recorded and/or transmitted, as shown in figure 2-6. It represents the aural analog of the old parlor stereopticon, in which each eye saw its own picture, the two having been taken by slightly separated cameras. If extreme care is taken in the modeling of the artificial head, then binaural listening over headphones can reproduce many of the fore-aft and height localization effects that we take for granted with normal listening.

In the normal listening environment, we make good use of slight head nodding, or *nutating*, to help us "zero in" on sound sources, and under such conditions our normal binaural acuity in lateral localization may be on the order of a handful of degrees. In the forward direction, lateral acuity is in the range of 3°, while at the sides it is in the range of 4.5°.

2.3.2   The Precedence Effect

If two loudspeakers produce the same signal at the same time, a listener equidistant from them will localize the apparent source of sound between the two. If one of the loudspeakers is delayed with respect to the other, then the listener will immediately localize at the earlier of the two loudspeakers. For delays up to about 5 msec, a 10-dB increase in the level of the

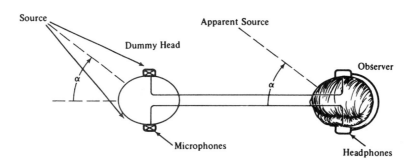

Figure 2-6. Binaural transmission.

delayed loudspeaker may be accommodated, with localization still tending toward the earlier loudspeaker. Data developed by Haas (1949) show the range of the delay-level trade-off, as shown in figure 2-7a.

Beyond about 5 msec, a constant 10-dB level differential can be accommodated, and this range extends to about 25 or 30 msec. Beyond that point, the listener begins to detect two distinct sound sources, and speech intelligibility may suffer.

Some useful things can be done with the precedence effect. In sound reinforcement, for example, loudspeakers located under balconies are normally delayed so that their sound arrives at the listener at the same time as (or slightly later than) that from the front loud-speakers. In that way, natural localization is preserved for those listeners.

Figure 2-7b shows an interesting trick using the precedence effect. Here, we have an undesirable echo from the back of a large room caused by a concave surface. Since the echo is some 60 msec behind the sound arriving from the front, it will be noticed as such and will be

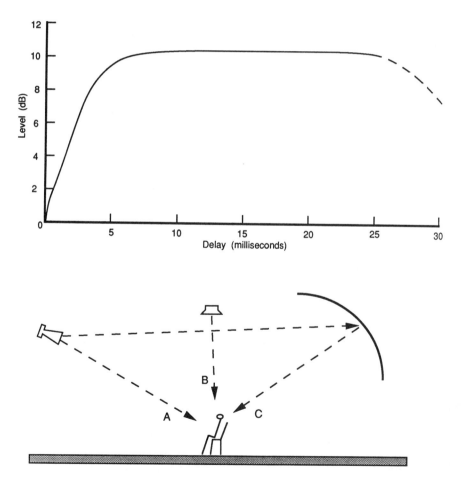

a

b

Figure 2-7. (a) The precedence effect; (b) echo suppression by means of the precedence effect. Sound A arrives first; sound C arrives 60 msec. later. Sound B is delayed so that it arrives at the listener 30 msec after sound A, thus masking sound C.

disturbing to listeners. Now, if a directional loudspeaker is placed over the echo interference area and is delayed at the listeners 30 msec with respect to sound arriving acoustically from the front, then we will have masked the echo from the back of the room.

What happens is that sound from the front of the room masks that from the overhead loudspeaker, because the delay is just within the allowable range. In the same manner, the sound from the overhead loudspeaker masks the echo from the rear. This is an example where "more is less."

2.3.3 Stereophonic Sound and Phantom Images

One of the earliest stereo demonstrations remains the most enlightening. Blumlein demonstrated in 1931 the recording scheme shown in figure 2-8. By calling this microphone array a "binaural pair," he demonstrated that he knew, perhaps intuitively, a great deal about how we hear.

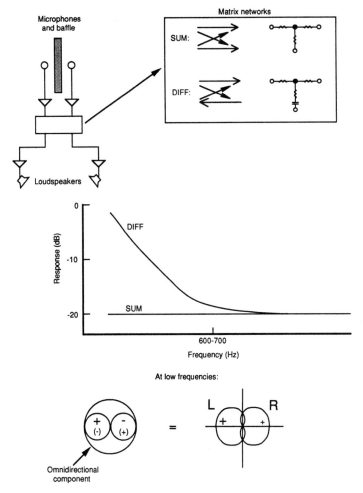

Figure 2-8. Details of Blumlein's "binaural pair" transmission system.

The spacing of the two omnidirectional microphones was roughly the same as that of the ears, and the padded baffle between them provided shadowing just as the head does. By matrixing and equalizing the microphones at low frequencies, he synthesized a pair of side-facing directional microphones. This was effective in the range of 600 or 700 Hz and below, and it provided ideal directional control in that frequency range. Above about 2 kHz, the interference effect of the baffle provided sufficient left-right channel differentiation in the recorded signals.

On playback over a pair of loudspeakers, accurate stereo perspectives were presented. Those sources of sound to the left of the microphone array were heard predominantly from the left loudspeaker, and the same held for the right. And for those listeners equidistant from the loudspeakers, sources of sound located toward the middle of the microphone array were heard along the median plane.

Such sound sources originating from positions where there are no loudspeakers are called *phantom images*, and they come about as shown in figure 2-9a. Note that a real sound source

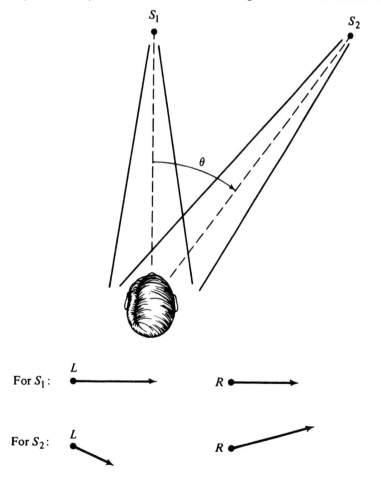

a

Figure 2-9. (*a*) Phasor relationships at the ears with real images; (*b*) phasor relationships at the ears with stereophonic loudspeakers.

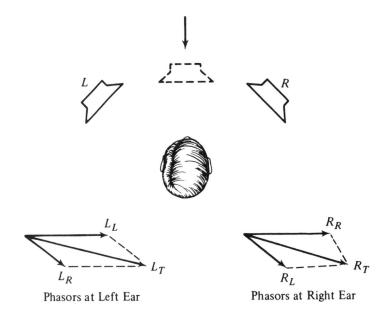

b Phasors at Left Ear   Phasors at Right Ear

located directly in front of the listener will produce equal phasors at both ears, and this is interpreted as a source directly ahead. A source displaced laterally will cause different phasors at the ears, with a leading angle at the nearer ear and a slightly louder signal at that ear.

The same conditions can be created by a pair of loudspeakers, and this is shown in figure 2-9b for a center phantom image. It is the summation of unequal phasors at each ear that produces net identical signals, thus giving the impression of a sound located directly ahead of the listener.

In a similar manner, sound sources located in front of the microphone array at intermediate positions will give rise to phantom images at those positions. It is even possible for phantom sound sources to be heard slightly outside the bounds of the playback loudspeaker array as a result of phasor summation at the ears of listeners exactly on the plane of symmetry.

Stereo, of course, has proceeded along many lines. We will discuss current recording techniques in a later chapter.

2.4 PITCH PERCEPTION

We normally associate pitch with frequency, but the two are not quite the same. Figure 2-10 shows the relationship between the mel and frequency. The *mel* is a measure of subjective pitch. Note that the curve is fairly linear up to a frequency of 500 Hz, but that above that frequency there is considerable divergence.

The data were gathered using pure tones played alternately, and an apparent 2-to-1 (octave) relationship represents a doubling of mels. At higher pitches, increments larger than a 2-to-1 frequency ratio are necessary to produce the sensation of octave relationships in pitch.

When pure tones are used, pitch is a function of level, as shown in figure 2-11. In general, pure tones of 1 kHz or lower tend to sound lower in pitch with increasing level, while those higher in frequency tend to sound higher in pitch with increasing level.

2.4.1   Absolute Pitch

A very few musicians possess what is called *absolute pitch* (AP). (The term *perfect pitch* is also used.) Absolute pitch is the ability to recognize the pitch of an isolated musical tone without reference to another tone. Trained musicians normally possess excellent relative pitch and can immediately identify a note once a reference has been set. The possessor of AP needs no such reference.

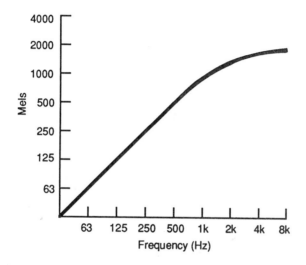

Figure 2-10. Subjective pitch (mels) versus frequency.

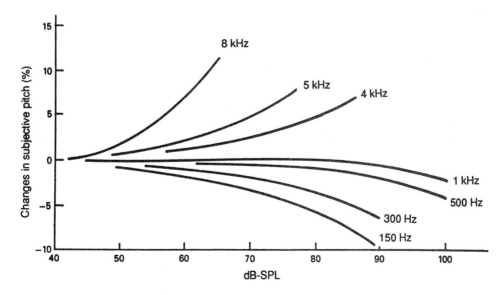

Figure 2-11. Dependence of subjective pitch on signal level.

The ability usually manifests itself during early musical instruction, and most evidence favors the theory that it is reinforced through early practice, rather than being simply an inherited ability. It is often associated with especially talented musicians, but it is certainly not a requisite for musical success at any level.

Possessors of AP make the same errors in octave judgment that other people do, and in standard psychological testing procedures they seem to hear the way other people do.

Some musicians possess what may be called quasi-AP. This is the ability to recognize a single pitch (C or A, for instance) and then let relative pitch take over from there. Some musicians have been so conditioned by their own instrument that they can "hear" in their mind's ear, without prior cues, certain tones that the instrument can produce.

Absolute pitch is not readily learned in adulthood, but the literature does point to a few successful cases (Ward and Burns 1982, 431).

2.5   CRITICAL BANDS; CONSONANCE AND DISSONANCE

In section 2.1, we discussed the place theory of hearing, describing the resolution of the cochlear response to pure tones as being that of a critical band. The measurement technique used here is shown in figure 2-12.

A fixed sine wave oscillator produces a tone f_1, and a variable oscillator produces a tone f_2. Over a given interval, shown in the figure, the two tones will fuse into a single tone whose frequency appears to be the average of the two. There will be slow beats in this region, but the tone's frequency will be clear.

At the upper and lower limits of this region of fusion, there will be zones of roughness, beginning at a point where there is a frequency difference of about 10 Hz. When the frequencies are separated further, the listener will begin to hear the two separate frequencies, and the roughness will disappear. The critical band is defined as the zone that bounds the roughness sensation and the central zone in which a single frequency is heard.

In a strict sense, the two closely separated tones are quite consonant, producing as they do only a slow beating sensation. The tones are once again consonant when they are separated sufficiently so that we hear both of them as separate entities. In between is a zone of dissonance, with its disagreeable roughness, as shown in figure 2-13. Over most of the range of musical fundamentals, the width of the critical band is very nearly the frequency ratio 6/5 (or about the interval of a minor third).

Another definition of the critical band is shown in figure 2-14. Here, a pure tone is set against a band-limited noise of equal power that just masks it. Stated another way, the power in the noise band is equal to the power in the pure tone, and the bandwidth of the noise signal has been adjusted so that it just masks the pure tone. The bandwidth of the noise signal is then defined as the critical band.

This measurement technique produces results nearly identical with those produced by the method shown in figure 2-12.

2.5.1   Cultural Conditioning

One person's consonance is another's dissonance, and this is based largely on cultural conditioning. The European tradition of diatonic composition has emphasized simple intervals, and relatively few of them within an octave, with clear-cut musical architecture. By

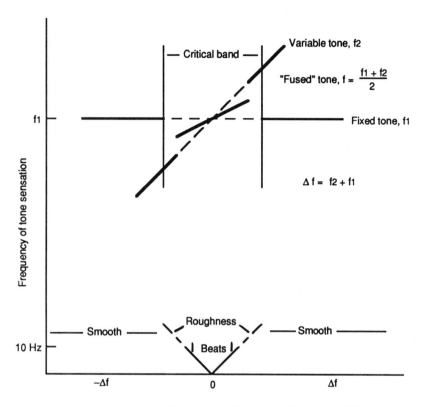

Figure 2-12. Determination of critical bandwidth. Data after Roederer 1973.

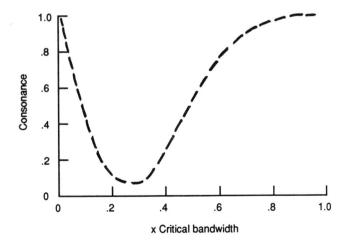

Figure 2-13. Consonance as a function of critical bandwidth.

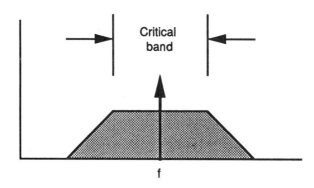

Figure 2-14. Critical bandwidth, another determination. Note: Band of noise and tone f have equal power.

comparison, some oriental cultures find such writing simplistic and devoid of subtlety. Similarly, the Western ear often hears oriental and Middle Eastern music as out of tune and structureless. Such is the nature of training, tradition, and time measured in millennia.

REFERENCES

Backus, J. 1969. *The Acoustical Foundations of Music*. New York: Norton & Co.

Bauer, B. 1956. "Phasor Analysis of Some Stereophonic Phenomena." *J. Acoustical Society of America* 33(11).

Bekesy, G. 1960. *Experiments in Hearing*. New York: McGraw-Hill.

Benson, K. 1988. *Audio Engineering Handbook*. New York: McGraw-Hill.

Beranek, L. 1962. *Music, Acoustics and Architecture*. New York: Wiley.

Berg, R., and D. Stork. 1982. *The Physics of Sound*. Englewood Cliffs, NJ: Prentice-Hall.

Blauert, J. 1983. *Spatial Hearing*. Cambridge, MA: MIT Press.

Blumlein, A., British Patent Specification Number 394,325. 1958. *J. Audio Engineering Society* 6(2) (reprint).

Dowling, W., and D. Harwood. 1986. *Music Cognition*. New York: Academic Press.

Gardner, M. 1973. "Some Single- and Multiple-Source Localization Effects." *J. Audio Engineering Society* 21(6).

Haas, H. 1949. "The Dependence of a Single Echo on the Audibility of Speech." *J. Audio Engineering Society* 20(2) (reprint).

Pierce, J. 1983. *The Science of Musical Sound*. New York: Scientific American Books.

Roederer, J. 1973. *Introduction to the Physics and Psychophysics of Music*. New York: Springer-Verlag.

Sandel, T., et al. 1955. "Localization of Sound from Single and Paired Sources." *J. Acoustical Society of America* 27:842–852.

Schubert, E., ed. 1979. *Psychological Acoustics*. Stroudsburg, PA: Dowden, Hutchinson, and Ross.

Stevens, S., and H. Davis. 1983. *Hearing, Its Psychology and Physiology*. Reprint. New York: American Institute of Physics.

Stevens, S., and J. Volkman. 1940. "The Relation of Pitch to Frequency: A Revised Scale." *American Journal of Psychology* 53:329–353.

Ward, W., and E. Burns. 1982. "Absolute Pitch." In *Psychology of Music*. New York: Academic Press.

Winckel, F. 1967. *Music, Sound and Sensation: A Modern Exposition*. New York: Dover Publications.

ADDITIONAL RESOURCES

Auditory Demonstrations. 1987. Compact disc available from Acoustical Society of America.

3

Scales, Temperament, and Tuning

Many persons involved in music technology have not been trained in music, and their knowledge of music theory is often limited. At the same time, they may have strong musical preferences, and possess "good ears," in the conventional sense of being able to discern musical nuances and fine pitch relationships. We will introduce some musical terminology in this chapter. Those readers with a basic musical background will have no difficulty, and for those without such a background, the definitions we give may bridge the gap. A few sessions at the keyboard with a musician should clarify the remaining points.

Today, we think of tuning of scales as something long standardized. Equal temperament, the division of the octave into twelve equal steps, is the basis of the tuning of all modern keyboard instruments, and the bulk of nineteenth- and twentieth-century music of European tradition was written with that tuning system in mind. But it has not always been so, and the history of tuning goes back to antiquity.

By way of definition, the term *scale* refers to an ordered series of notes that provides a basis for tonal composition. Such terms as major, minor, whole tone, and so forth, describe scales. The term *temperament* refers to the precise interval relationships between the notes of a scale, and the term *tuning* strictly refers to the actual process of adjusting the pitch of a given vibrating element. In general usage, however, the terms *temperament* and *tuning* have become nearly synonymous.

In this chapter we will examine the rich history of temperament, noting the events that led to the general acceptance of equal temperament. We will also discuss some of the factors that have led to a new interest in some of the older tunings. Finally, we will discuss pitch standards and the notion of key chroma.

3.1   MAJOR AND MINOR SCALES

Figure 3-1*a* shows a one-octave segment of the keyboard spanning the notes from C to the C above. This constitutes the scale of C major. A major scale can begin on any of the twelve notes shown, with all intervals suitably transposed up or down. However, throughout this chapter all discussions will be restricted to C major and its related A minor scale. One form of the A minor scale is shown in figure 3-1*b*.

The scales of C major and A minor, and all similar transpositions, are referred to as *diatonic*, meaning that they all have a "key center." In traditional harmony of the eighteenth and nineteenth centuries, this implied a system of chords, and functional relationships between chords, that provided the harmonic foundation of musical composition.

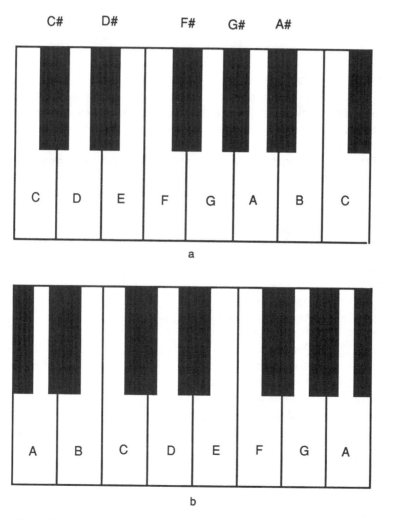

Figure 3-1. View of scales on the keyboard: (*a*) C major; (*b*) natural minor scale based on A.

3.1.1 Scale Steps and Intervals

In order to define functional harmonic relationships, specific nomenclature was developed, and the steps of the diatonic scale were given names to indicate their basic musical functions, as shown below:

Scale Step	Name
C	Tonic
D	Supertonic
E	Mediant
F	Subdominant
G	Dominant
A	Submediant
B	Leading tone
C	Tonic

In this terminology, *tonic* implies the tonality, or the "home base" of the key. *Dominant* implies the importance of the function that resolves directly to the tonic step. Thus, a chord built on the dominant step will normally resolve to a chord built on the tonic or key step, thus defining the key center. The leading tone, in a melodic context, usually resolves upward, or "leads," directly to the tonic. The remaining terms, with their sub- and super- prefixes, define the positions of those steps in the scale.

Another method of identifying the positions of notes in the scale is by interval, as shown below:

Interval	Nomenclature
C–C	Unison
C–C-sharp	Augmented unison
C–D-flat	Minor second
C–D	Major second
C–D-sharp	Augmented second
C–E-flat	Minor third
C–E	Major third
C–F	Perfect fourth
C–F-sharp	Augmented fourth
C–G-flat	Diminished fifth
C–G	Perfect fifth
C–G-sharp	Augmented fifth
C–A-flat	Minor sixth
C–A	Major sixth
C–A-sharp	Augmented sixth
C–B-flat	Minor seventh
C–B	Major seventh
C–C'	Octave

Certain intervals, such as C–D-sharp and C–E-flat, are composed of the same notes on the keyboard and of course sound the same. They are termed *enharmonic*, and the difference in interval "spelling" is based on the specific musical functions the intervals perform.

The complete scale, including all adjacent black and white notes, is called the *chromatic* scale. The interval between two adjacent notes in the chromatic scale is referred to as a *semitone*, while the interval made up of two semitones is called a *tone*.

3.2 THE HARMONIC SERIES

When a vibrating string is plucked or bowed, it can undergo a variety of motions. Figure 3-2a shows the simplest motion, corresponding to the definition of resonance that we gave in chapter 1. Here, the string moves not as a unit, but with a single overall motion.

The same string can also be made to execute the motion shown in figure 3-2b. In this case, the string is moving as if it had a boundary at the middle, creating two strings of half length. Similarly, if properly excited, the string can vibrate in thirds, as shown in figure 3-2c.

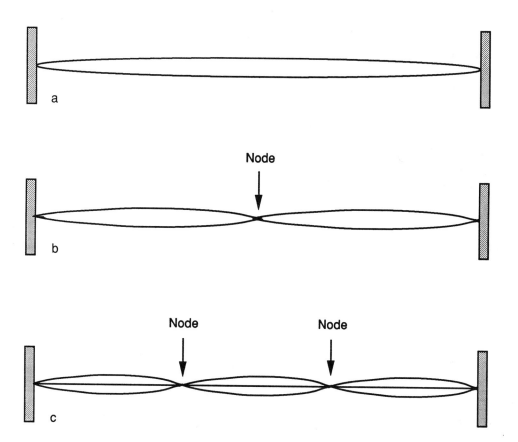

Figure 3-2. Vibration of a string (*a*) at the fundamental; (*b*) at the second harmonic; (*c*) at the third harmonic.

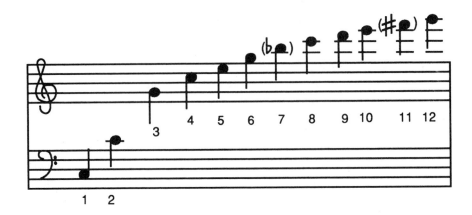

Figure 3-3. The harmonic series based on C₃; first 12 harmonics shown.

This process can extend well upward, and all higher-frequency modes of vibration are called *harmonics*. The *harmonic series* is the entire set of motions that the moving system can resonate to, and the ratios of frequencies are all based on simple integers. Thus, the fifth harmonic has five times the frequency of the fundamental, the seventh harmonic seven times, and so forth.

Figure 3-3 shows the first twelve notes of the harmonic series built on the note C below middle C on the piano keyboard. The fundamental frequency is 130.8 Hz, and all the harmonics are integral multiples of this frequency.

Intervals that are fairly low in the harmonic series sound quite pleasing and "pure" to the ear, and in the early years of scale theory and evolution these intervals were used in scale construction. Certain higher harmonics (shown in parentheses) sound out of tune and have not generally been used in scale construction.

As examples, the frequency ratio between the third and second harmonics is 3/2, and that between the seventh and fourth is 7/4. The definition is straightforward and will be used as we construct scales.

3.3   THE PYTHAGOREAN SCALE

This scale dates from antiquity and is based on intervals of the fourth and fifth. The C major scale is given below:

Note:	C	D	E	F	G	A	B	C
Frequency Ratio:	1/1	9/8	81/64	4/3	3/2	27/16	243/128	2/1
Interval Ratio:		9/8	9/8	256/243	9/8	9/8	9/8	256/243

For example, note that the interval between C and D is 9/8, the same as the corresponding interval between C and D, the ninth and eighth harmonics, as shown in figure 3-3. The interval between D and E is the same 9/8 ratio. However, if we multiply these intervals to arrive at the C-to-E frequency ratio, we get 81/64.

Note, however, that in the harmonic series the frequency ratio between C and E is 5/4, which corresponds to a frequency ratio of 80/64. The interval between the so-called natural major third (derived from the third and fifth harmonics) and the Pythagorean major third (derived from successive perfect fifths) is (81/64) × (64/80), or 81/80.

This ratio between the natural major third and the Pythagorean major third, called the *syntonic comma*, played an important role in scale evolution in the early years. (The term *comma* implies a tuning error or adjustment, in this case a relatively small one.)

The remaining notes in the Pythagorean scale, represented by the black keys in figure 3-1, were derived by simply extending the basic interval relationships. For example, C-sharp was defined as a 9/8 ratio relative to the note B, with the other black notes similarly defined.

Generally speaking, the Pythagorean scale was quite useful for early music, which did not stray far from the "home key" in which the basic temperament was generated. The fifths and fourths were for the most part pure, and this was important for certain music.

Please remember that the scale we have analyzed could just as easily have been constructed on any of the twelve notes of the keyboard, in a sense making the scale twelve times transposable.

3.4   THE JUST SCALE

Another early temperament, this scale was based entirely on simple intervals in the harmonic series. Its construction is based on the 6 : 5 : 4 interval ratio of the major triads built on F, C, and G. The ratios between C, G, and F were established as perfect fifths. The scale is given below:

Note:	C	D	E	F	G	A	B	C
Frequency Ratio:	1/1	9/8	5/4	4/3	3/2	5/3	15/8	2/1
Interval Ratio:		9/8	10/9	16/15	9/8	10/9	9/8	16/15

Like the Pythagorean scale, the just scale was suited to music that did not stray far from its home tuning center. But as music developed and the interval of the third took on more importance, a family of temperaments known as meantone were developed.

3.5   MEANTONE TEMPERAMENT

The term *meantone* implies an averaging of some kind. In the case of this tuning, the averaging is in the string of fifths: C–G–D–A–E. This extended series of intervals gives the ratio (3/2)(3/2)(3/2)(3/2), which is equal to 81/16. When this is doubled back to the major third interval of C to E, the ratio becomes (81/16)(1/4), or 81/64.

In the tuning process, this interval is reduced to 80/64, or 5/4, by an amount equal to the syntonic comma discussed earlier, and the "error" is equally distributed among the notes in the string of fifths. The resulting scale has a just major third between C and E, with fifths that were only slightly detuned from the ideal 3/2 ratio. Similarly, just major thirds were maintained between D and F-sharp, F and A, and G and B.

With meantone tuning, a new family of temperaments arose, some of which are very useful today in the performance of early keyboard music.

3.6   EQUAL TEMPERAMENT

It was realized early that there was no simple temperament that would work ideally in all keys. However, a consensus developed that, if modulation were to be freely treated in composition, equal temperament would ultimately be adopted. In equal temperament, all twelve notes of the scale shown in figure 3-1 are divided into equal frequency ratios. Since there are twelve of them, and since the octave ratio is 2 to 1, the ratio between individual notes becomes the ratio of the twelfth root of 2, or

$$\text{Frequency ratio} = 2^{1/12}$$

This ratio is 1.05946. If we multiply the frequency of C by this ratio, we get the frequency of C-sharp. Continuing the process through the twelve notes of the chromatic scale, we will arrive at the interval of 2 : 1, which of course defines the octave.

None of the intervals derived by this process will be pure. For example, in equal temperament, the perfect fifth, C to G, will have a ratio of 1.4983. By comparison, the just interval is 1.5, or 3/2. The equally tempered perfect fifth is slightly flat compared with the just interval, but generally this is not enough to bother musicians.

The major third in equal temperament is given by the ratio of 1.2599. By comparison, the just major third has a ratio of 1.25. Here, the error is not so easily ignored, and it is generally felt today that the major third is the most "out of tune" of all intervals in equal temperament. For the most part, modern ears have become accustomed to it and freely accept it.

3.6.1   The Cent

As a way of keeping track of ratios between intervals in equal temperament, or any other system of tuning, the *cent* is a useful concept. It is a logarithmic unit, like the decibel, and there are 100 cents in each equally tempered semitone. The octave thus contains 1200 cents, inasmuch as there are 12 semitones per octave.

The basic relationship is given as

$$\text{Cents} = \frac{(1200)(\log R)}{\log 2}$$

$$\text{Cents} = (3986)(\log R) \tag{3.1}$$

where R is the interval frequency ratio and logarithms are taken to the base 10.

Thus, for a ratio of 3/2, which is the ratio of the just perfect fifth,

$$\text{Cents} = (3986)(\log 1.5)$$

$$\text{Cents} = (3986)(0.176) = 702 \text{ cents}$$

By comparison, the tempered perfect fifth consists of seven semitones, and the number of cents is 700, so the tempered perfect fifth is just slightly smaller than the just perfect fifth.

Taking into account the fact that a tempered quarter tone is 50 cents, an error of up to 10 or 15 cents may be considered negligible in most musical contexts.

Earlier we mentioned that the tempered major third was the most "out of tune" interval in the equal tempered system. The interval is equal to 400 cents. By comparison, the just major third is equal to

$$\text{Cents} = (3986)(\log 1.25)$$
$$= 386 \text{ cents}$$

Thus, the error between the two is 14 cents, noticeable in some musical contexts, but negligible in others.

The Pythagorean major third, an interval with a ratio of 81/64, can also be expressed in cents:

$$\text{Cents} = (3986)(\log 81/64)$$
$$= 407 \text{ cents}$$

The Pythagorean major third is even wider (sharper) than the tempered major third, and may thus be more annoying in certain musical contexts.

3.7   EQUAL TEMPERAMENT IN MUSICAL PRACTICE

Equal temperament is the "glue" that holds modern music performance together. However, at any given time, certain instruments may deviate from it to good effect. For example, a vocal chorus or the string ensemble of a symphony orchestra, with their ability to sing or play any pitch in their range, will often place leading tones slightly sharp in order to make a more natural resolution to the tonic one-half step above. Likewise, at a final cadence, the major third above the tonic may be flattened slightly to lend a stronger feeling of repose.

As we shall see in a later chapter, the normal "pitch fringe" in ensemble performance covers a multitude of slight mistunings, giving an overall effect that is quite satisfactory musically and that makes so much of the fuss about tuning and temperament seem quite academic and beside the point.

The ability of a musician to "play in tune" is a measure of how well that player or singer hears a note and places it in musical context. The term *intonation* is used to describe this precise placement of pitch.

A pianist, of course, has no intonation problems—unless the instrument is out of tune. A violinist has a wide range of pitch control limited only by natural hearing ability and technical skill. A woodwind or brass player can coax a tone up or down by a slight amount, but only with some possible compromise in playing dexterity.

3.8   OTHER TUNINGS IN CONTEMPORARY PERFORMANCE

Without considering exotic scales from the Orient, a good bit of modern musical performance can benefit from using some of the older temperaments. Arthur Benade (1976, 312–313) describes such a situation:

It should be clearly understood that the problems of producing good music on sustained-tone instruments having notes of fixed pitch are by no means insuperable, although they do pose a challenge to the skill and imagination of the composer. The following anecdote will perhaps help you to understand the situation. I recently had occasion to hear a concert which I tape recorded for acoustical reasons with the permission of the performer. The program included the following three compositions for pipe organ: Toccata per l'Elevatione from the Messa delli Apostoli (1635) by Girolamo Frescobaldi; Prelude and Fugue in A major (BWV 536) by J. S. Bach; and Etude I (1967) by Gyorgy Ligeti. The first of these pieces contained a number of sustained major thirds, which work perfectly well on an organ tuned to one of the unequal temperaments common in the seventeenth century, but which fight unmercifully on today's equally tempered instruments. During the playing of it the audience stirred uneasily, and, when I have played the tape, numerous musicians (including pianists and harpsichordists) have asked me what terrible thing went wrong with the organ. Most are incredulous when the explanation is given, even when they listen to the piece by Bach played the same on the identical organ, sounding in the Bach like the admirable instrument it is. Bach . . . arranged for his thirds to come and go, well disguised by their musical context. Even close listening does not bring out the roughnesses that we know are present in the bare interval. The third composition, a modern one, directly exploits the roughness of equally tempered thirds. One long-held chord follows another without let-up in slow and hypnotic progression of changing registration, pitches, and beating intervals.

Through the decade of the eighties, a number of pipe organs have been built with modified meantone temperaments that favor common keys. In a liturgical context, some keys are rarely encountered, while others are very routine. A temperament that takes this into account may, overall, perform a musical service to the listeners and to the choral performers.

3.9   PITCH STANDARDS

In most of the world, A–440 Hz is the reference for tuning. In some locations, orchestras tune slightly higher than this, often with the idea that a slightly brighter sound will result from it. There is a clear penalty placed on the woodwind and brass players when this approach is overdone, since the instruments may not be capable of precise intonation under such conditions.

Most eighteenth-century music performance on period instruments is based on A–415 Hz, a good half tone lower than current standards. This was, it is assumed, the standard of the day, and many of the string instruments of the period could be tuned to the modern standard only with some difficulty.

Organ tunings varied even more, and some of the extreme variations in pitch standard are shown in Table 3-1.

Considering that all these standards were tunings for the same written note, the spread seems excessive. Euler's clavichord, for example, sounded exactly a whole tone lower than today's instruments, while the pitch of J. S. Bach's organ in Hamburg was eventually a whole tone sharp. Mersenne's pitch was more than a major third higher than today's pitch standard.

The situation is somewhat reminiscent of the early days of mechanical recording, where the nominal standard of 78 rotations per minute was only an approximation. The actual recording speed was set using mechanical governors with weight-driven turntables. Speeds were set

Table 3-1. Pitch Standards over the Years

Ensemble or Instrument	Pitch for A Above Middle C, Hz	Cents re A–440 Hz
Euler's clavichord (1739)	392	−200
Pascal, court tuner, Paris 1783	409	−126
Dresden, Silbermann organ (1754)	415	−101
Praetorius' "suitable pitch" (1619)	424	−64
Scheibler's Stuttgart Standard (1834)	440	0
Steinway, America (1879)	457	+66
Jakobikirche, Hamburg (1688)	489	+183
Jakobikirche, Hamburg (1879)	494	+200
Mersenne, ton de chambre (1636)	563	+427

with stopwatches, and there were no exact standards. Knowing the key of a piece of music, one could always adjust the playback governor to match the pitch of a piano known to be in tune. But even then one could not be absolutely sure.

Today's early music practice seeks out not only original instruments but original tunings, to the extent that the information is available.

3.9.1   Frequency Meters

A competent piano tuner does not rely on a frequency meter, inasmuch as no meter can do as well as practiced ears (although some tuners will store the temperament for a given instrument in a sophisticated meter, in order to facilitate later retuning of the same instrument). There are, however, certain important applications for tuning meters. Musicians in live performance situations must tune various string instruments, such as harps and guitars, while there is much noise around them. Under these conditions, even a skilled ear has a difficult time hearing the pitch accurately, and a tuning meter is a handy thing to have. Such meters are available for about one hundred dollars and are quite accurate for such applications.

3.10   THE PITCH SPIRAL; KEY CHROMA

In the mind's eye, pitch is related to a spiral, with each loop representing one octave, as shown in figure 3-4. A given note—C, for example—always appears at the same location on the spiral, with all its octaves directly above and below. In calling out isolated frequencies (once a reference has been set), many musicians will make octave errors, and think nothing about it. It is more fundamental to identify the correct key center than to identify a note as being in the second or third octave above middle C.

Some musicians think in terms of key chroma; that is, they associate a particular subjective quality, mood, or other attribute with a given key. The term *chroma* comes from the Greek word for color, and in this context we are using it in a more general sense, but noting as well that some musicians actually associate certain colors with certain keys.

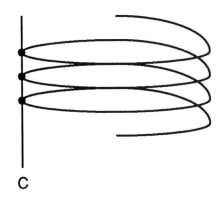

C

Figure 3-4. The pitch spiral.

As we have seen, early temperaments did allow each key to have its own chroma through a unique set of ratios between intervals. A keen ear can sometimes spot these and, through fancy or imagination, assign qualities to a given key.

With equal temperament most of these cues go away, at least for keyboard instruments. However, the string group, with its open string tunings on C, G, D, A, and E, may take on one kind of quality when playing in the keys that emphasize these open strings, especially on tonic and dominant steps of the scale. When playing in keys that make more use of stopped strings, the sound is likely to be a bit softer and more mellow, inasmuch as the sound of an open string is quite different from that of a stopped string.

In the woodwind section of the orchestra, breaks between registers of the instruments create tricky fingerings in certain keys. Not all keys are created equal on all instruments, and musicians have always been aware of this.

Most musicians will state that their feelings for and responses to certain keys, whatever they may be, go back to their early training in childhood, a time of great susceptibility to analogies, qualitative attributes, and the detailed characteristics of the instruments on which they learned to play.

REFERENCES

Backus, J. 1969. *The Acoustical Foundations of Music*. New York: W. W. Norton.

Barbour, J. 1953. *Tuning and Temperament*. East Lansing, MI: Michigan State College Press.

Benade, A. 1976. *Fundamentals of Musical Acoustics*. New York: Oxford University Press. (Quote by permission of Virginia Benade.)

Berg, R., and D. Stork. 1982. *The Physics of Sound*. Englewood Cliffs, NJ: Prentice-Hall.

Culver, C. 1956. *Musical Acoustics*. New York: McGraw-Hill.

Pierce, J. 1983. *The Science of Musical Sound*. New York: Scientific American Books.

ADDITIONAL RESOURCES

The New Oxford Companion to Music, New York: Oxford University Press, 1983.

4

Musical Sound Generation and Radiation

In this chapter we will study the physical bases of traditional musical instruments, observing how power is applied to the instrument, how oscillations are produced and altered, and how useful sound radiation is achieved. We will treat these elements in the simplest possible way, reserving for later chapters the detailed discussions of specific instruments, their unique characteristics and fine differences.

Curt Sachs (1940) divides traditional musical instruments into four major categories:

Chordophones: These are instruments that employ strings under tension; methods of excitation include plucking, bowing, and striking. Typical instruments in this category include the orchestral strings, guitar, harp, and piano.

Aerophones: These are instruments that employ an air column, driven by a suitable source, to produce and sustain sound. Included here are all brass instruments and the woodwind families of reeds and flutes.

Membranophones: These are instruments that employ stretched membranes that are struck. They may produce sounds of definite or indefinite pitch. Drums of all types are in this category.

Idiophones: These are instruments that are self-resonant, producing sounds when struck or otherwise excited mechanically. The group includes bells, cymbals, wood blocks, and the like.

There are subfamilies within each group, and there are some instruments that defy easy categorization. The free reed of a harmonica, for example, is classed as an aerophone, but it has no air column. Certain primitive instruments are spun, producing a "whirring" sound.

strictly speaking, they are aerophones, inasmuch as the moving cords or leather thongs do not emit sound, but give rise to audible air turbulence. The tambourine, of course, is both a membranophone and an idiophone, and the piano has many characteristics of a percussion instrument.

4.1 BASIC POWER RELATIONSHIPS

All traditional instruments have four physical elements in common: a source of power, a means of coupling power to the instrument, an oscillating element or system, and some means of coupling useful acoustical power, or sound, to the surroundings.

Direct mechanical power produced by the player is the most common method of exciting instruments. All percussion and string instruments are sounded in this manner. The aerophones are generally excited pneumatically by the performer's breath, although the bagpipe is powered by a combination of breath and applied force by the player's upper arm.

Methods of coupling power to instruments are quite varied: hands, feet, hammers, sticks, bows, mouthpieces, and reeds, to name a few.

The oscillating element is usually that part of the instrument by which it is assigned to its basic category. With the idiophones, there is usually one note per oscillating element, and if more than one note is required, there is a set of such elements, normally arranged chromatically. The xylophone and glockenspiel fall under this description.

Some string instruments, such as the harpsichord and piano, similarly have a string or group of strings for each note. The violin family, on the other hand, relies on lengthening or shortening strings by stopping them against the fingerboard with the left hand. On the orchestral harp, there are seven strings in each octave, forming the diatonic scale of C. Seven pedals are used to raise or lower each octave set of strings by one semitone.

The aerophones conveniently lengthen or shorten the oscillating air column using valves, slides, or keyed openings on the side of the air column. Changes of the player's lip control, or *embouchure*, provide additional alteration of the tuning of the air column.

Finally, useful sound radiation is achieved through efficient coupling of the vibrating system to the air medium. Strings are nearly always attached or coupled to a sounding board. Relatively large in area, the sounding board receives mechanical motion from the strings and provides better acoustical coupling to the air than can the string alone, with its very small surface area. The piano has a relatively large sounding board, while that of the violin is rather small. Lower-frequency radiation from the violin is aided by acoustical resonance in the cavity and coupled outward through the openings in the body of the instrument. At the very highest frequencies, the vibration of the bridge of the violin contributes significantly.

Most idiophones couple directly to the air, since they are large enough to radiate efficiently at their natural frequencies. Some idiophones, such as the marimba and vibraphone, utilize resonant air columns beneath each bar for more efficient radiation. In this sense, these instruments may be thought of as hybrids between idiophones and aerophones.

The aerophones as a group are acoustically well matched to the air. The bells on woodwind instruments have little to do with increasing the radiation capability, since most of the radiation takes place through the open key holes. In the case of brass instruments, the bell provides many functions. Primarily, it couples the vibration in the pipe effectively to the air and helps align the vibrational modes of the pipe into a useful set of harmonics. Secondarily, it increases the directivity of higher harmonics and gives the player better control in the placement of higher pitches.

Specific details of sound radiation for the various instrumental groups will be covered in chapters 5 through 9.

4.2   THE CHORDOPHONES—STRINGS AND SOUNDING BOARDS

Figure 4-1 shows a string under tension, coupled to a sounding board by means of a bridge. In the seventeenth century, Mersenne (1636) determined the mathematical factors that govern the pitch of a string:

1. Frequency is inversely proportional to the length of the string.

2. Frequency is proportional to the square root of the tension in the string.

3. Frequency is inversely proportional to the thickness of the string.

4. Frequency is inversely proportional to the square root of the string's density.

Mersenne's conditions can be reduced to a relatively simple equation, and if the string is set into motion at its midpoint, it will vibrate at a fundamental frequency given by

$$f = \left(\frac{1}{2L}\right) \sqrt{\frac{T}{m}} \qquad (4.1)$$

where L is the length of the string (meters), T is the tension of the string (newtons), and m is the mass per unit length of the string (kilograms/meter).

Many readers will note the similarity between this equation of motion and equations 1.6, 1.7, and 1.8. The basic physical relationships are essentially the same.

Figure 4-2a shows the simple motion of the string as given by equation 4.1. If we stop the string in the middle and excite it at a point one-quarter the distance from one end, the motion shown in figure 4-2b will result.

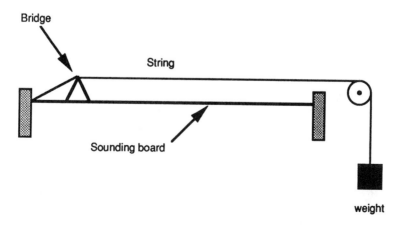

Figure 4-1. A string under tension.

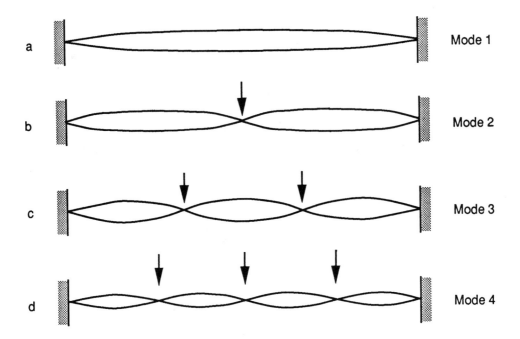

Figure 4-2. Vibration of a string in integrally related modes.

In a similar manner, the string can be made to vibrate in thirds, fourths, fifths, and higher modes. These various vibrations are referred to as mode 1, mode 2, mode 3, and so forth. In the general case, the string can vibrate in all such modes at the same time, giving rise to complex waveforms. In the simpler modes of vibration shown in figure 4-2, those positions on the string where there is no motion are called *nodes* and are indicated by small arrows. The positions of maximum vibration are called *antinodes*.

4.2.1   Standing Waves

What we are observing in the vibrating string are *standing waves*. A disturbance along the string travels from one end to the other; an upward deflection reflects as a downward deflection, and vice versa. The result of this action is the appearance of a stationary or standing wave along the string.

It is obvious that a vibrating string will always have nodes at each end, inasmuch as these points are fixed and cannot be displaced. It is also obvious that a string cannot have a node at the point of excitation, inasmuch as the string is being forced into motion at that point. It should also be clear that excitation of a string toward one end will give rise to many harmonics, inasmuch as many high-frequency antinodal points exist close to the ends of the string.

4.2.2   Driving the String

So far, we have only plucked the string, giving rise to a damped oscillation. If we bow the string, we can produce continuous oscillation. Figure 4-3 shows the effect of bowing; as the bow is drawn across the string, friction builds up sideways tension in the string, eventually reaching a point at which the string slips back to its original position. The process begins anew, and the string is thus maintained in steady vibration.

In the case of the piano, the strings are struck by felted hammers located a short distance from one end, while in the harpsichord the string is plucked by a small piece of hard leather or crow's quill, or perhaps plastic in modern instruments. The player's fingers and fingernails are used for plucking the strings of guitars, lutes, and related instruments.

4.2.3 Coupling the Strings to the Sounding Board

The sounding board draws mechanical power from the vibrating string, and the termination on the string at the bridge can be described as "lossy," as shown in figure 4-4. The greater the coupling between the string and the sounding board, the greater the damping in the system.

In most instruments the string runs more or less parallel to the sounding board, and one would expect the power transfer to be greatest when the string moves in and out with respect to the sounding board. However, in actual playing, the string may be excited in a direction parallel to the board. The actual movement of the string, once it is set into motion, is elliptical, and it can excite the sounding board through a combination of in-out motion and rocking motion of the bridge.

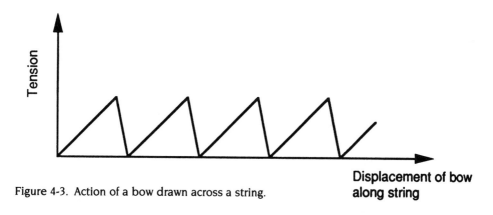

Figure 4-3. Action of a bow drawn across a string.

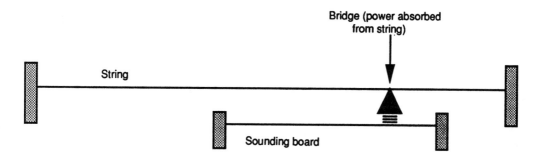

Figure 4-4. Transfer of acoustical power from a string to a sounding board.

In the harp, the strings make a large angle with the sounding board, and the transfer of power is not quite as efficient as in those cases where the sounding board and the strings are parallel. This is especially true at lower frequencies, and the output from the instrument is weak in that range.

4.3   AEROPHONES I—A SIMPLE OPEN AIR COLUMN

A column of air behaves in a manner similar to a string. As shown in figure 4-5a, a pipe that is open at both ends will resonate with antinodes at each end and a node in the middle. In this case, the node represents an air pressure maximum and an air particle velocity minimum. The antinodes represent air particle velocity maxima and pressure minima.

A standing wave is set up in the pipe, with velocity antinodes at each open end. For the fundamental (mode 1), there is one-half cycle of the waveform, as shown.

The frequency for mode 1 is given by

$$f = \frac{c}{2L} \tag{4.2}$$

where c is the velocity of sound (meters) and L is the length of the pipe (meters).

As shown in figure 4-5b and c, a pipe that is open at both ends can support all integral multiples of the fundamentals. Here, the second and third harmonics are shown. The flute is acoustically open at both ends and follows the description given here.

4.3.1   The Stopped Pipe

A pipe that is stopped at one end will produce the pattern of nodes and antinodes shown in figure 4-6. The fundamental mode of vibration supports only one-quarter wavelength, and a stopped pipe therefore has a fundamental an octave lower than an open pipe of the same

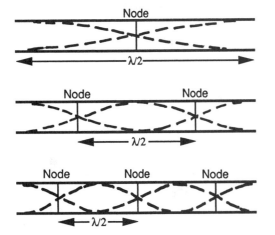

Figure 4-5. Mode structure in a pipe open at both ends.

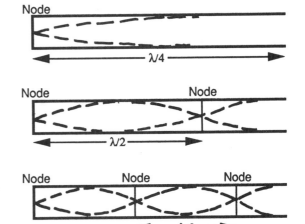

Figure 4-6. Mode structure in a pipe closed at one end.

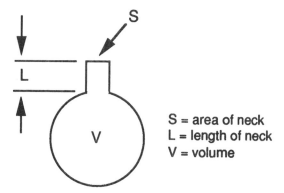

Figure 4-7. The Helmholtz resonator. S = area of neck; L = length of neck; V = volume.

length. Furthermore, the harmonics present in the stopped pipe are all odd (1, 3, 5, and so forth), since the pattern of antinodes at each end, required for the even harmonics, cannot be set up.

Obviously a stopped pipe cannot be keyed, since the key openings would render it an open pipe. The so-called slide flute, often used for special effects, is the only representative of the genre normally encountered in modern music. The pan flute, or "pipes of pan," consists of a diatonic set of stopped flutes that the player excites by blowing across the top of each pipe.

4.3.2 Exciting the Air Column; Bernoulli's Principle

Bernoulli's principle states that in a fluid system, an increase in velocity will be accompanied by a drop in pressure. An example given is the simple case of blowing across the narrow neck of a bottle. The air velocity across the opening gives rise to low pressure in the neck of the bottle, drawing air out of it. This then sets the "air spring" inside into oscillation with the air mass in the neck, and sustained oscillation results.

This is the principle of the Helmholtz resonator, which is shown in figure 4-7. The equation of resonance is

$$f = \frac{c}{2\pi} \sqrt{\frac{S}{LV}} \tag{4.3}$$

where S is the cross-sectional area of the neck (square meters), L is the length of the neck (meters), and V is the internal volume (cubic meters).

Note again the familiar form of the equation of resonance.

The same principle is used in playing the flute; the player's lips form a small air jet that is directed at an opening just slightly displaced from the end of the instrument. The term *embouchure* may be used to describe the exact position of the player's lips relative to the mouthpiece of the instrument.

In the case of the "penny whistle" or recorder flute, there is a fixed mouthpiece that directs the jet of air against a sharp edge known as a fipple. Figure 4-8 shows the two ways in which open or stopped pipes can be excited. In figure 4-8a, the flute player's lips form the jet directly. In figure 4-8b, we see the fixed embouchure as used in flageolets and in all flue organ pipes.

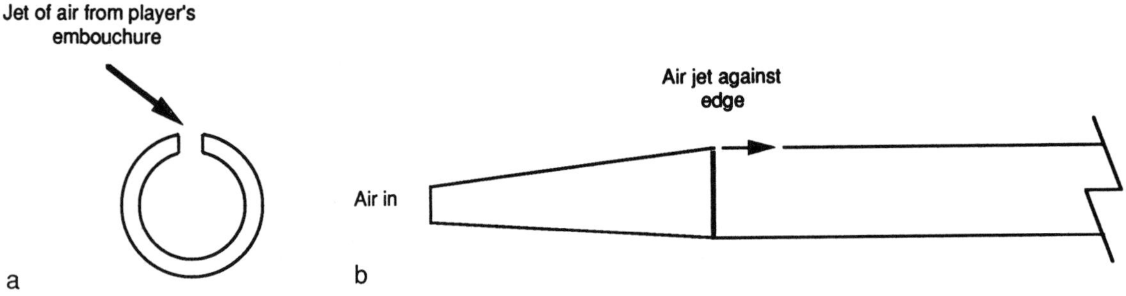

Jet of air from player's
embouchure

Air jet against
edge

Air in

a   b

Figure 4-8. Two ways of exciting oscillations in a pipe: (a) lip embouchure; (b) fixed embouchure.

4.3.3   End Corrections

The acoustical length of an open or stopped pipe is slightly longer than its physical length, inasmuch as the antinode at the radiating end exists at a small distance from the physical end of the pipe. Thus, the pitch of a pipe will be slightly lower than that calculated using its physical length. For a stopped pipe, with only one radiating end, the correction is given by the following equation:

$$L' = L + 0.58R \tag{4.4}$$

where L' is the effective length of the pipe, L is the physical length of the pipe, and R is the radius of the pipe.

For a pipe open at both ends, the correction is twice the amount given by equation 4.4.

4.3.4   Playing Higher Notes

The normal keying system on the flute provides slightly more than one octave's range by progressively shortening the speaking length of the pipe. If the player overblows and alters the air jet accordingly, the instrument will resonate at the second harmonic, and the playing range of the instrument will be increased. If the player wishes to reach even higher registers, a combination of overblowing to produce higher harmonics, reshaping the air jet, and opening certain upper keys may be used.

A similar technique can be employed in organ pipes, as shown in figure 4-9. The "harmonic flute" is an organ stop with pipes that are twice their speaking length. A hole located at the middle of the pipe coaxes the pipe into sounding an octave higher after it has first attempted to speak at the natural pitch of its physical length. The rise of one octave is accomplished by forcing an antinode at the center of the air column.

4.4   AEROPHONES II—REEDS AND RESONATORS

Here we will describe the basic sound production in instruments such as the clarinet, the saxophone, and the double reeds. When a reed is placed at one end of a cylindrical or conical resonator, the reed end always acts as a velocity node, and the open end always acts as a

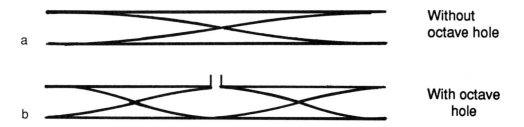

Figure 4-9. Exciting the second mode (octave) of an open pipe by a hole midway in the pipe.

velocity antinode. The resonators are either basically cylindrical, as in the case of the clarinet, or conical, as in the case of the saxophone and double reed families.

4.4.1   Single Reed with Cylindrical Resonator

This combination represents the basis of the clarinet family, which is the only orchestral wind instrument that behaves substantially like the stopped pipe discussed in section 4.3.1. Figure 4-10 shows the basic operation. When the player applies air pressure, the following events take place:

    a. Bernoulli's principle causes the reed to close. There is an abrupt compression that travels the length of the pipe.

    b. When the compression reaches the end of the pipe, the discontinuity in radiation impedance causes it to be reflected back as a pressure rarefaction. When it reaches the mouthpiece, the pressure rarefaction maintains the closed position of the reed.

    c. The rarefaction returns once again to the open end of the pipe, and the reed opens by its own oscillation.

    d. The impedance discontinuity at the open end of the pipe causes a reflection, this time as a compression, and the cycle begins anew.

This series of events has required four passes, or two "round trips," through the pipe. Thus, the pipe will have a physical length that is one-fourth the wavelength generated, and in this manner resembles the stopped pipe.

As in the case of the flute, we can force the cylindrical pipe with reed into the next higher mode of operation. Since the system can sustain only odd harmonics, we must place a hole at a point one-third the distance from the reed, thus forcing a velocity antinode at that point, as shown in figure 4-11. When the register key is opened, the note produced by the clarinet will be an octave and a fifth higher, corresponding to the third harmonic of the fundamental.

4.4.2   Single Reed with Conical Resonator

The conical resonator performs in a quite different manner from the cylindrical one. It produces a set of modes, as shown in figure 4-12. The lowest usable mode, shown in figure

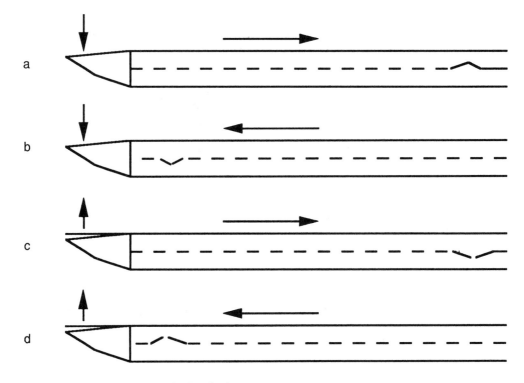

Figure 4-10. Wave travel in a cylindrical tube.

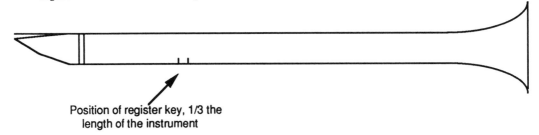

Position of register key, 1/3 the
length of the instrument

Figure 4-11. Position of the register key in a clarinet.

4-12a, is one octave higher than in the case of the cylindrical resonator, and the family of modes includes both odd and even harmonics. The second and third modes are shown in figure 4-12b and c. The reason for this fundamental difference between the conical resonator and the cylindrical resonator is that, because of its conical shape, the resonator does not have as abrupt an impedance discontinuity at the open end. There is little energy reflected from the open end back toward the reed, and this forces a redistribution of nodes and antinodes through the length of the resonator.

The conical reed instruments include the saxophone family and the double reeds (oboe and bassoon). They are provided with octave keys, in some cases more than one, in order to facilitate playing in upper registers.

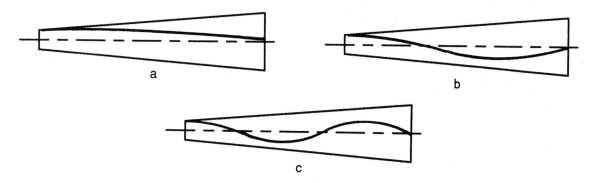

Figure 4-12. The first three modes of vibration in a conical reed system.

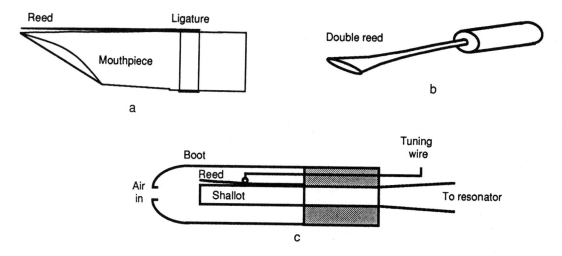

Figure 4-13. Three kinds of reeds: (*a*) single reed; (*b*) double reed; (*c*) organ reed assembly.

4.4.3   Reed Mechanisms

Figure 4-13 shows three reed mechanisms. The single reed is shown in figure 4-13*a*. Note that the reed lies against a tapered part of the mouthpiece. The application of wind pressure forces the reed to vibrate. Of itself, the reed-mouthpiece combination can vibrate over a wide range of frequencies. When it is coupled to the resonator, the resonator dominates, locking the reed's motion to that of the air column. A reed cannot vibrate at higher than its natural frequency.

   The double reed (figure 4-13*b*), as used in the oboe and bassoon, is constructed of two sections of cane placed against each other. The sections are relatively small and stiff, requiring

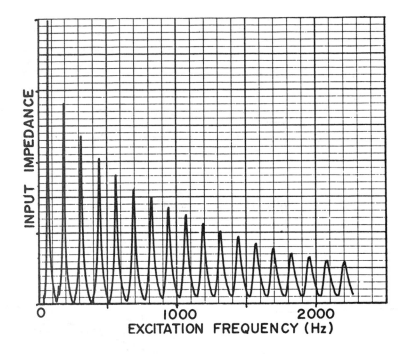

a

Figure 4-14. Impedance curves of various brass combinations: (a) cylindrical tube without bell; (b) cylindrical tube with bell; (c) trumpet with mouthpiece. From *Fundamentals of Musical Acoustics*, by Arthur Benade, copyright 1976. Reprinted by permission of Virginia Benade.

higher wind pressure on the part of the player than do the single reeds. The reed assembly is engaged between the player's lips, which are drawn tightly over the teeth.

The organ reed mechanism (figure 4-13c) works only at one frequency. It is tuned to the desired frequency by altering its length, and the resonator is then attached and tuned to the same wavelength, forming a coupled resonant system.

4.5   AEROPHONES III—THE LIP REEDS

The lips, plus mouthpiece, tube, and bell, form the basis for the brass family of instruments. The principle is ancient, going back to the early use of hollow animal horns as a means of signaling. The analysis is similar to that for the reed instruments.

4.5.1   The Cylindrical Resonator With and Without Bell

A cylindrical resonator will respond to a pair of vibrating lips the way it responds to a reed, producing the family of odd harmonics shown in figure 4-14a. Again, as in the case of the reed family, the addition of a relatively large, flaring bell will cause the system's resonances to shift, as shown in figure 4-14b. When a mouthpiece is added, the pattern of resonances is further modified, as shown in figure 4-14c. Resonances in the mouthpiece itself cause the system to be more responsive in the frequency range from about 250 to 1200 Hz, and this makes it possible for the player to excite the system fairly easily at a number of higher harmonics.

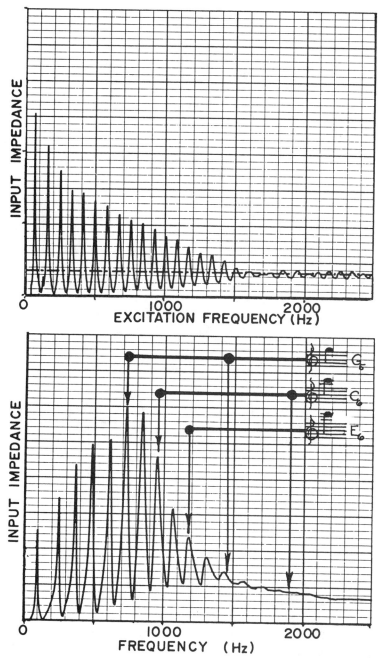

b

c

When a harmonic series of notes exist, the frequency ratios correspond to the numbers 1, 2, 3, 4, . . . , n. Any note in the series can itself become a new fundamental. For instance, let us take the second harmonic. The ratio of notes 2, 4, 6, 8, and so forth, forms a new harmonic series, since their ratios can be reduced to 1, 2, 3, and 4. Likewise, a series may be built on the third harmonic: 3, 6, 9, 12, and so forth.

What this means is that a player can "lock in" on any note in the first few vibrational modes of the air column and produce that tone clearly as a new fundamental. In the case of the bugle, which is a simple brass instrument without valves or other means of changing speaking length, the player can, through embouchure (lip) control, play notes corresponding to modes 3, 4, 5, and 6. This simple group of four notes provides the tonal basis for all the traditional military bugle calls. As a group, all the brass instruments are capable of being driven at high harmonics, while the reed instruments are normally not capable of playing accurately and easily higher than their second or third harmonics.

4.5.2   The Conical Resonator with Bell

The conical resonator used in some brass instruments represents only a slight change in cross-sectional area with respect to length, as compared with the cylindrical resonator. The physical factors are essentially the same, and both types of resonators produce the same harmonic series. The cylindrical form generally tends to produce more harmonic development than the conical form. Both are fitted with rapidly flaring bells at the open end. We should note that the profile of most brass instruments is quite narrow relative to overall length. A trumpet has a diameter of about 1 cm (0.4 in.) over most of its length, and the length itself is about 160 cm (63 in.). By comparison, a clarinet has a diameter of about 2 cm (0.8 in.) and a length of about 60 cm (24 in.). The relatively narrow bore of brass air columns, plus the action of mouthpiece resonance, makes it easy for those instruments to produce new fundamentals fairly high in their harmonic series.

4.5.3   Mouthpieces

Figure 4-15 shows typical brass instrument mouthpieces. The trumpet mouthpiece (figure 4-15a) is quite shallow, and its small air chamber facilitates the production of higher harmonics. The French horn mouthpiece (figure 4-15b) is about twice as deep, contributing to a more mellow tone with less harmonic development. The trombone mouthpiece (figure 4-15c) is essentially a larger version of the trumpet mouthpiece, scaled appropriately for the frequency range of the instrument.

Players are very sensitive to extremely small changes in the profile and bore of a mouthpiece, and a mouthpiece that is ideal for one player may not suit another.

4.6   MEMBRANOPHONES

A stretched membrane when struck will execute complex motion composed of a number of specific modes. Calfskin and plastic are the materials normally used. The modal structure of the vibrating membrane, or drumhead, is complicated by the fact that it is a two-dimensional surface. We must therefore define two variables, radial modes and circular modes (and combinations of the two), as illustrated in figure 4-16. The two-digit number at the top of each illustration indicates the modal combination, the first indicating the radial modes and the second indicating the circular modes. Frequency ratios relative to the 01 mode combination are given below each illustration.

Mode structure 01 indicates that the drumhead is moving in and out uniformly, while mode structure 11 indicates a rocking motion of the drumhead. It is clear that the modal combina-

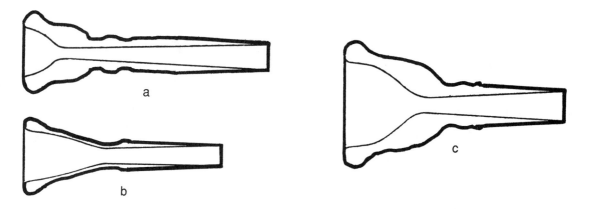

Figure 4-15. Profiles of various brass mouthpieces: (a) trumpet; (b) French horn; (c) trombone.

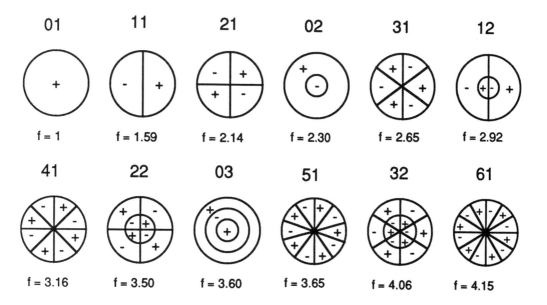

Figure 4-16. Theoretical modes for a circular diaphragm.

tions are not harmonically related, and this means that it is difficult to assign a clear pitch to the vibration.

The modal structure shown here may be modified considerably by the nature of the shell to which the head is attached. In the case of the timpani, which we will discuss in a later chapter, the modifications clearly produce a definite pitch. In the case of a drum with a membrane on each rim of the shell, the mode structure is influenced by the enclosed air mass.

The relative frequency of mode structure 01 depends on the tension in the head as well as the mass per unit area of the head.

4.7   IDIOPHONES

Because of their complex shape, many idiophones nearly defy analysis; there are, however, some simple forms, such as bars, rods, tubes, and plates, that are fairly easily analyzed.

As a single example, we show in figure 4-17 the first four bending modes of a bar. Note that the modes of vibration are not integrally related.

Many idiophones vibrate in a very complex manner in which torsional (twisting) modes may combine with bending modes. The degree of damping may vary between mode types, giving rise to unusual tonal characteristics. The manner of striking the idiophone profoundly affects which modes will be dominant.

In a later chapter, we will continue the discussion of specific idiophones.

4.8   STARTING AND STOPPING TRANSIENTS

The onset of sound in any instrument has a characteristic transient, or "start-up," quality that is not part of the continuous tone. For example, the action of the bow on a string will cause a momentary wide band noise as the bow excites the string into oscillation. In the aerophones, the starting transient is usually caused by the abrupt increase in air pressure against the reed or mouthpiece, and the transient may consist of upper harmonics of the resonant system. In the case of the various percussion instruments, the transient usually consists of higher-pitched modes and low-frequency thumps, both of which usually damp out fairly quickly. While such sounds are not normally thought of in a musical context, music devoid of them would sound very strange indeed. Similarly, the stopping of tone may be accompanied by a characteristic transient, normally less noticeable than that at the start.

4.9   THE VOICE

A physical view of the vocal tract is shown in figure 4-18a, and a functional diagram (after Benade) is shown in figure 4-18b. Air is forced by the lungs through the larynx, where the interrupting effect of the vocal chords produces a "spiky" sawtooth signal. The pitch of this signal is determined by muscle-controlled tension in the vocal chords, and the frequency range varies from about 90 Hz to about 1 kHz, depending on the type of voice and gender.

Vowel sounds are created by formants, or specific resonances in the mouth and nasal cavities. Such sounds as *ah*, *oh*, and *ee* are characterized by certain resonance frequencies, and these frequencies, or formants, are superimposed on the output from the vocal chords.

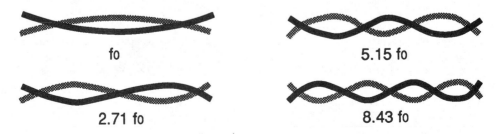

fo   5.15 fo

2.71 fo   8.43 fo

Figure 4-17. First four bending modes of a bar.

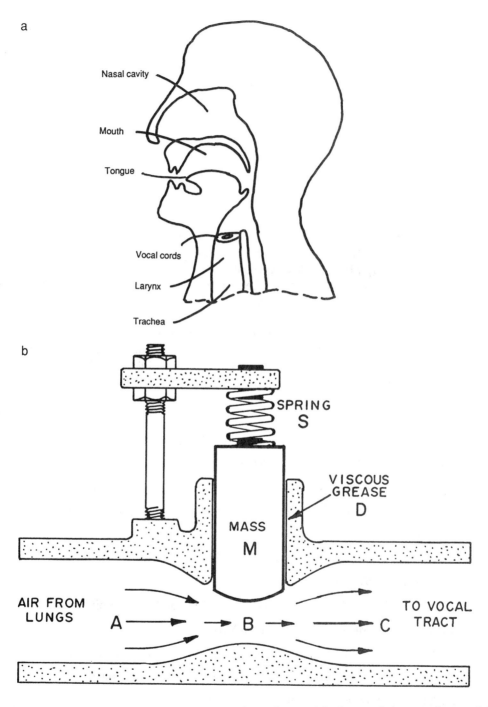

Figure 4-18. Details of the voice mechanism. (*a*) Physical view of the larynx; (*b*) functional view of the larynx. From *Fundamentals of Musical Acoustics*, by Arthur Benade, copyright 1976. Reprinted by permission of Virginia Benade.

Table 4-1. Approximate Vowel Formant
Frequencies

Vowel Sound	Dominant Frequencies
oh as in "hole"	500, 850
o as in "law"	600, 1100, 2900
oo as in "pool"	300, 700, 2500
ah as in "father"	600, 1100, 2600
ee as in "need"	200, 2500, 3200
eh as in "pet"	550, 1900
ay as in "say"	550, 2100

In addition to the steady-state vowel sounds, there are a number of transient sounds, such as those indicated by the letters *s, t, k, p, ch, f, m,* and *n,* that are produced in the upper part of the vocal tract, largely by specific placement of the tongue and lips.

Speech consists of meaningful inflection of the basic signal generated in the larynx, along with formation of vowels and placement of the various transient sounds. In the singing voice, inflection is indicated by the pitch of the written note and is often accompanied by a vibrato, or gentle modulation of frequency in the 2- to 5-Hz range.

Formants may vary slightly between singing and speaking, and of course spoken formants may vary as a result of cultural or regional differences. Table 4-1 shows the formant frequencies for the basic vowel sounds.

REFERENCES

Backus, J. 1969. *The Acoustical Foundations of Music.* New York: W. W. Norton.

Benade, A. 1976. *Fundamentals of Musical Acoustics.* New York: Oxford University Press.

Berg, R., and D. Stork. 1982. *The Physics of Sound.* Englewood Cliffs, NJ: Prentice-Hall.

Campbell, M., and C. Greated. 1987. *The Musician's Guide to Acoustics.* New York: Schirmer Books.

Culver, C. 1956. *Musical Acoustics.* New York: McGraw-Hill.

Mersenne, M. 1636. *Harmonie Universelle.*

Moravcsik, M. 1987. *Musical Sound.* New York: Paragon House Publishers.

Olson, H. 1952. *Musical Engineering.* New York: McGraw-Hill.

Pierce, J. 1983. *The Science of Musical Sound.* New York: W. H. Freeman.

Rossing, T. 1990. *The Science of Sound.* Reading, MA: Addison-Wesley.

Sachs, C. 1940. *The History of Musical Instruments.* New York: W. W. Norton.

5

Acoustics of String Instruments

This is the first of five sequential chapters in which we will discuss the major instrument groups. While chapter 4 provided the basis for sound production, these chapters will fill in details on instrument construction, basic playing techniques, and other characteristics unique to the individual instruments. We will also present information gathered from many sources regarding the spectral, dynamic range, and directional performance of the members of the various instrument groups. The aim here is to fill in many gaps in the reader's knowledge of instruments and to approach instruments from the point of view of the player. A strengthening of knowledge in these areas will benefit the musical technician in a most important way.

5.1   THE VIOLIN

The violin of today is little changed from the instruments made by Nicolo Amati (1649–1740), Antonio Stradivari (1644–1737), and others of the Cremona school, which flourished in Italy through the first half of the eighteenth century. The violin is shown in figure 5-1a and b. In external appearance it looks bilaterally symmetrical, but internally it is not.

The vibrating strings are coupled to the body of the instrument via the bridge. The strings are set in motion by bowing or plucking, and the speaking length of the strings is determined by left-hand finger position on the fingerboard. Open string tuning is accomplished by pegs that adjust the strings to the desired tension.

While the instrument appears delicate, it is relatively robust, considering its many fine dimensions. The downward force of the tuned strings on the bridge is about 8 kg (17.6 lb) weight, and the tension in the direction of the strings is about 22 kg (48.4 lb) weight. These considerable forces are balanced by the arched shape of the top plate and by the back via the internal sound post.

Materials have always been part of the mystique of the instrument. Although the type of wood may vary, spruce is normally used for the front plate (belly) and back plate of the instrument. The back plate is often constructed from mirrored sections of wood and appears

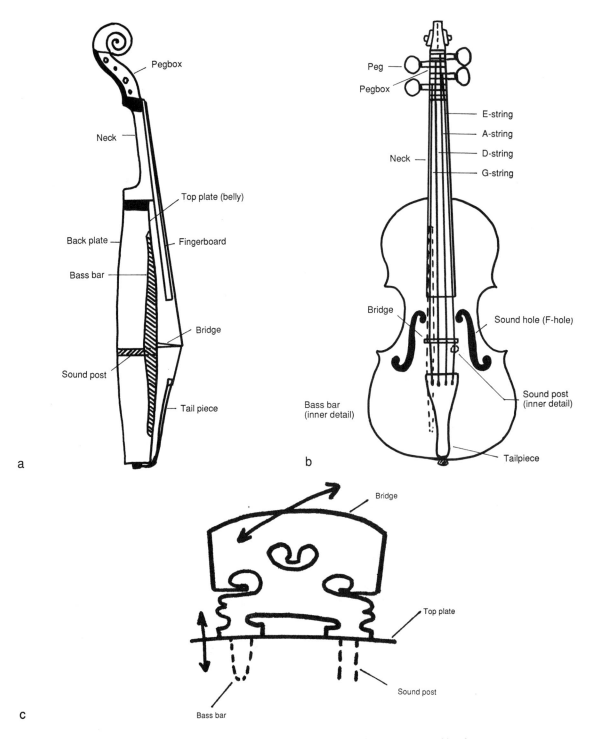

Figure 5-1. (*a*) External view of the violin; (*b*) internal details; (*c*) motion of bridge.

symmetrical about the centerline of the instrument. The thickness is about 3 mm (0.1 in.) in the center, and somewhat thinner at the edges of the plates.

The ribbing, which separates back and front, is usually made of maple, as are the neck and bridge. The fingerboard and tailpiece are made of ebony. The acoustical effects of the many kinds of varnishes that have been used in finishing the instrument are not clearly understood.

Physically, the bridge is positioned high above the top plate so that motion of the bow on the outer strings can easily clear the body. The "waists" in the sides of the body evolved with this requirement in mind. The motion of the bridge is largely influenced by the positioning of the soundpost, a round piece of pine or spruce that is wedged internally between the front and back plates close to the treble foot of the bridge. In this position, the soundpost acts as a fulcrum, allowing the bridge to rock up and down about its bass foot and thus transfer motion to the top plate, as shown in figure 5-1c. Since the soundpost is wedged in place, it can be adjusted by a violin maker, and this can change the instrument's performance significantly.

The bass bar, which is made of pine or spruce, is glued to the underside of the top plate just under the bass foot of the bridge (G-string side). Its purpose is to distribute the rocking motion of the bridge uniformly over the top plate.

Output at lower frequencies (G_3 to C_5) results from complex motion of the front and back as well as from air resonances coupling through the "f-holes" in the top plate. The main air (Helmholtz) resonance frequency is about 290 Hz, which is quite close to D_4, the natural tuning of the second string of the instrument. The lowest significant body, or wood, resonance is quite close to 440 Hz, thus reinforcing the output from the third string of the instrument (A_4).

In some instruments "wolf notes" may be a problem. These are interferences between certain plate resonances and the air resonance in the body. The result is a note that differs from its neighbors in quality or loudness. Beats can also occur between the fingered note and an especially sharp resonance, creating intonation problems. A good violin maker can often minimize the problem through repositioning of the soundpost.

In the middle range of the instrument (above C_5), most of the sound is radiated by complex motion of the top plate of the instrument. At higher frequencies, the bridge becomes progressively decoupled from the top plate, eventually assuming most of the radiation load itself at the highest harmonics.

In the middle and high ranges, no two instruments will radiate in quite the same way, inasmuch as the complex modes of vibration of body and bridge are unique to each instrument.

Thus, we may visualize the instrument as somewhat akin to a high fidelity loudspeaker system, with distinct woofer, mid-range, and tweeter elements.

While the body of the instrument has remained fairly constant over the years, the neck and fingerboard of the instrument have been rebuilt and repositioned, as shown in figure 5-2, in order to elevate the bridge and permit the use of greater string tension. The change is in keeping with the instrument's musical evolution into its virtuoso role of today. Note in particular the longer fingerboard in the modern instrument. This modification extended the range of the instrument considerably at higher frequencies.

5.1.1   The Strings and Bow

Traditionally, strings were made of catgut, with the low (G_3) string wound with wire. Today, lamb gut is largely used, and the second and third strings (D_4 and A_4) are wound with

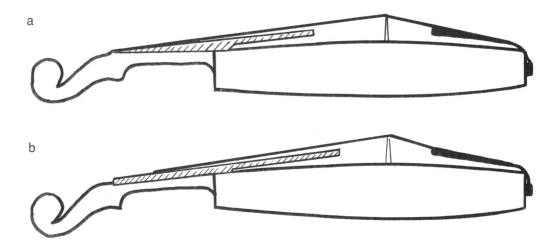

Figure 5-2. Neck, bridge, and body relationships (a) in older violin construction and (b) in newer construction.

Figure 5-3. The bow. (a) Shape of old bows and (b) shape of new bows.

aluminum. The highest string (E_5) is normally made of steel. Wirewound strings provide added mass so that the desired tuning can be accomplished without reduced tension. This enables the player to produce more acoustical output with the desired degree of control. The steel E string gives considerable brilliance to the upper registers of the instrument.

The primary "stick and slip" action of the bow was described in chapter 4. In performance, the mechanical behavior of the bow is crucial, and one can spend well up in the thousands of dollars for a good one.

As the instrument has evolved, so has the bow. Figure 5-3 illustrates the basic differences between the early bow and the modern one developed by François Tourte (1747–1835). The modern bow is curved in such a way that the application of greater force against the string increases the tension in the horsehair ribbon itself, resulting in relatively little displacement of the ribbon. By contrast, greater force on the older bow displaces the ribbon considerably, making it difficult to play loudly without engaging two or more strings at once. Thus, the modern bow allows the player to exert greater force for louder playing, while maintaining good articulation and control of the instrument. The modern bow is about 75 cm (30 in) long and is quite light. Its center of gravity is about 20 cm (8 in) from the end held by the player. The best examples of bows are made of pernambuco wood from South America.

5.1.2   Playing Techniques

The violin is not supported entirely by the player's left hand, as may be commonly thought. The instrument is largely supported between the chin rest and the player's left clavicle (collarbone). The left hand is then free to move up or down the fingerboard with the necessary agility.

Bowing is the normal method of producing sound, and there are numerous techniques:

1. *Legato* (connected) bowing: normal smooth up-and-down motion of the bow, with a direction change on each note. Normal playing with the bow is sometimes indicated *arco* in musical scores.

2. Detached bowing: each note given a firm down bow motion (*détaché*, *martellato*).

3. Bouncing of the bow (*spiccato*).

4. Playing with the bow close to the bridge to produce a sound very rich in harmonics (*sul ponticello*).

5. Tapping the strings with the wood of the bow (*col legno*).

6. *Tremolo*: a fast back-and-forth motion of the bow.

7. Sliding between notes (*portamento*).

Legato playing is normally done with a *vibrato*, which is produced by a back-and-forth motion of the left hand that gives a slight pitch undulation above and below the note. The vibrato rate varies among players, but is generally in the range of 4 to 6 Hz.

Harmonics can be played on the instrument by lightly touching the strings at a nodal point and bowing at an antinodal point. Natural harmonics are played on open strings, while artificial harmonics are played on stopped strings. In this latter case, the string is stopped by the first (index) finger, and another finger touches the string at the desired nodal point.

Double stopping is the playing of two notes at once. The technique requires great skill.

Scordatura is the intentional mistuning of open strings for special effect. Normally, the detuning would be to lower the open pitch of a string in order to extend the compass of the instrument downward.

Pizzicato refers to plucking of the strings. This may be done by either hand, as required.

The instrument may be muted by wedging a small wooden or metal piece onto the bridge, fitting between the strings. The added mass of the mute diminishes the sound output, creating a somewhat nasal sound.

5.2   THE STRING FAMILY

There are four instruments that make up the orchestral string section, and they are all related structurally to the violin. The respective open string tunings of these instruments are shown in figure 5-4. These instruments are shown in the upper section of figure 5-5.

The scaling of these lower-pitched instruments does not follow that of the violin. The viola, in particular, is relatively small for its normal range of frequencies. Its lowest string is a perfect fifth below that of the violin, but its overall body length is in the range of 38 to 44 cm (15 to 17 in) as opposed to the 35.5-cm (14-in) body length of the violin. Its relatively small size permits it to be played in the same manner as the violin.

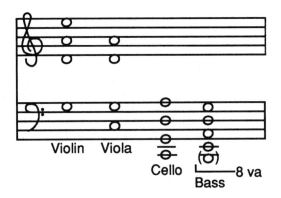

Figure 5-4. Open string tunings for the bowed orchestral string instruments.

The cello (or violoncello) and bass viol are played in a different manner, resting on the floor with the neck pointing upwards. The body length of the cello is about 75.5 cm (30 in), slightly less than twice the length of the violin. The bass viol (also called double bass or contrabass) has a body length of about 114 cm (45 in). Total tension along the strings of the cello is in the range of 50 kg (110 lb) weight, while that of the bass viol is in the range of 170 kg (374 lb) weight.

Note that the bass viol is tuned in fourths, while the other members of the family are tuned in fifths. On the three higher-pitched instruments, the player's four fingers can easily span the five diatonic notes contained within the interval of a fifth without changing hand position. With the bass viol, the distance along the string is too great for this to be the case, at least for someone with hands of normal size; thus, it is necessary to tune the instrument in fourths.

The bass viol is often provided with a C extension. This is an apparatus that is attached to the low E string, carrying its speaking length down to C_1. The low string is normally stopped at E_1, and a set of four levers lengthens the string chromatically down to C_1. Some players prefer to use a fifth string tuned to B_0 in order to reach lower notes. The bass part is written one octave higher than it sounds.

In jazz and popular music, the bass viol is normally fitted with an electrical contact pickup on the bridge so that it can be amplified.

5.2.1 Distribution of Primary Resonances in the String Family

As we stated earlier, the violin is characterized by its Helmholtz resonance at about the frequency of its second string (D_4) and its primary body resonance at about the frequency of its third string (A_4).

The viola cannot be so neatly categorized, inasmuch as the size of the instrument may vary. Generally, the Helmholtz resonance is around 230 Hz and the primary body resonance around 350 Hz. As a result, the lowest notes on the instrument seem weak by comparison with the corresponding range on the violin. The cello is pitched an octave lower than the viola, and its Helmholtz and primary body resonances are even higher in relation to the open string tunings than those of the viola.

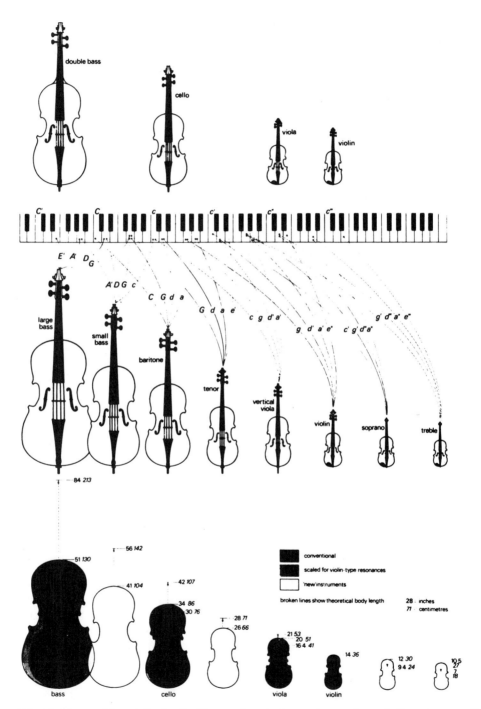

Figure 5-5. Members of the violin family. The modern instruments are shown in the upper portion, while the "new violin family" is shown in the middle section. A comparison of body sizes is shown at the bottom. Data courtesy of the Catgut Acoustical Society.

Thus, the lower members of the group are not simply "bigger violins." As we have seen, their evolution has been shaped largely by physical and performance limitations. Furthermore, the distribution of body and Helmholtz resonances gives the string section of an orchestra its characteristic spectrum, in which the bulk of the energy is concentrated in the range from G_2 to D_4.

5.2.2   A New Family of Violins

In the 1950s, Carleen Hutchins and Frederick Saunders (1875–1963) developed a true family of violins, as shown in the middle portion of figure 5-5; they have their Helmholtz and main body resonances aligned with their two middle strings, as the violin does. Generally, this has called for bodies much larger than those of the traditional instruments, and in the case of the alto instrument (corresponding to the viola), it is easier for many players to play the model in cello fashion, with the foot of the instrument resting on the floor.

Several composers have written for the ensemble of new violins, and the effect is strikingly different from that of the standard instruments. Their use in a traditional string quartet ensemble would certainly lead to balance problems, but there is no questioning the application of the alto and baritone models in place of the viola and cello for sonatas and concertos.

5.3   THE GUITAR

Instruments such as the guitar, banjo, mandolin, and a host of folk instruments are the latter-day descendants of the lute family. Superficially, these instruments resemble the violin, with body, neck, strings, and bridge. An important difference, however, is the flat bridge and top plate and the use of frets along the fingerboard. The frets are raised strips of metal placed crosswise along the fingerboard that allow the player to stop a string at a precise point. In the modern guitar, nineteen frets are located on the fingerboard at chromatic intervals, giving the player a range of an octave and a fifth on each string.

The so-called classical, or Spanish, guitar is shown in figure 5-6a. Structural details are shown in figure 5-6b, and the tuning of the six open strings is shown in figure 5-6c. Today, the strings are usually made of nylon, and the lower three are often wound with fine wire.

The classical guitar is normally plucked with the fingernails, but the fleshy ends of the fingers may be used as well. Additional playing techniques include strumming (sliding the thumb across the strings), playing with the fingernails close to the bridge (creating a metallic sound), and the production of harmonics. In the Spanish folk idiom, certain percussive effects are common on the instrument. These are produced by tapping the fingernails of the right hand against the body of the instrument. A familiar attachment for the guitar is the *capo*, a device that can be strapped to the fingerboard at any fret position, enabling the instrument to be played at higher transpositions without any change in fingering.

Like instruments of the violin family, the guitar has evolved with certain relationships between Helmholtz and body resonances. The Helmholtz resonance is around 100 Hz, corresponding to the note $G\#_2$. The main body resonance is about one octave higher. As in the case of the violin, the various modes of the table or top plate are quite complex, and no two instruments radiate sound in precisely the same way.

In jazz and popular ensembles, the guitar is normally outfitted with an electrical pickup so that its sound can be amplified.

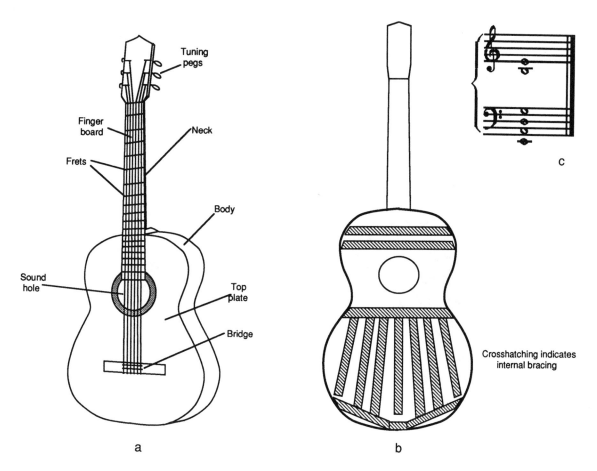

Figure 5-6. The classical guitar. (*a*) The instrument; (*b*) construction details; (*c*) open string tunings.

5.4   THE HARP

Like instruments of the guitar family, the harp has seen many forms and played a large role in the development of folk music worldwide. The modern orchestral instrument may be referred to as the pedal, or diatonic, harp, inasmuch as it has seven pedals, each of which controls the pitch of one of the seven steps of the diatonic scale. The mechanism was patented by Erard in 1810.

Figure 5-7*a* shows a view of the harp. There are seven strings per octave. For ease in identifying the various octaves, all the C strings are red and the F strings are blue.

Each pedal has three positions. The middle position is the natural one, and when the pedals are set in this position, the harp can produce the notes of the C-major diatonic scale. If, for example, the E pedal is raised one notch, then all the E strings in the instrument will be lowered to sound E-flat. If the pedal is lowered one notch, all the E strings will sound E-sharp,

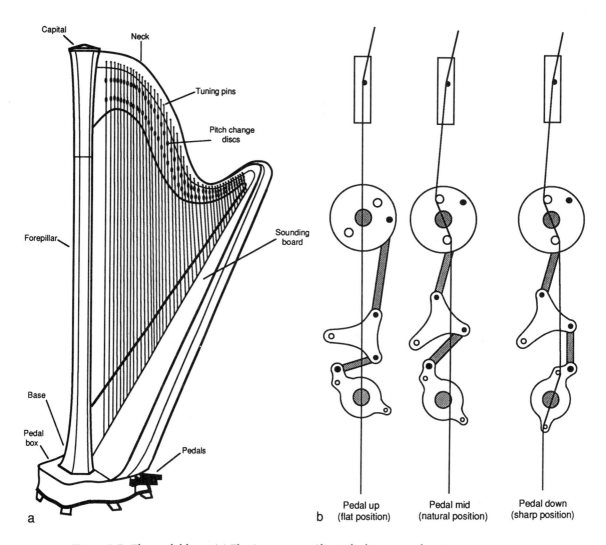

Figure 5-7. The pedal harp. (*a*) The instrument; (*b*) pitch change mechanism.

which is the enharmonic equivalent of F. A detail of the pitch-changing mechanism is shown in figure 5-7*b*.

Thus, the pedals may be set initially to play in a given key, with changes being made as required to allow for chromatic passages and various harmonic combinations. Skilled players can make fairly rapid pedal changes, but arrangers and composers are urged not to overtax the players in this regard. In writing or making arrangements for the instrument, it is necessary to be aware of how the mechanism works and, more to the point, what it cannot do. Certain chords are not possible on the instrument. If such sonorities are demanded in orchestral writing, then two or more instruments must be used.

Only the thumb and first three fingers are used in playing the harp. Each hand can cover the entire compass, but the left hand usually plays in the lower register of the instrument. Normally, chords are rolled (sounded from the bottom upward) unless otherwise indicated. The *glissando* is a special harp effect in which the thumb, or the fingernail, is drawn rapidly up or down the strings, producing a great sweep of sound.

The strings may be plucked at various positions, giving a wide variety of tone. Harmonics are played by placing the base of the thumb at the nodal point and plucking with the end of the thumb.

The harp string will sound until it dies out naturally, unless the player damps it by pressing the palm against the still sounding string. But the decay rate is fairly rapid, as compared with the piano's, so there is relatively little overlapping of undamped sounds.

5.5   MODERN VERSIONS OF THE BASS AND THE GUITAR

Many modern rock and popular ensembles make use of electric basses and guitars. These are solid body instruments that make no use of body or air resonances, as in the conventional instruments. Only the strings resonate, and the relatively undamped tones have great sustaining ability. The tunings of these instruments follow those of their predecessors.

The output from these instruments is entirely through electromagnetic pickup and subsequent amplification. In the electronic path there may be many special effects, including reverberation, equalization, and dynamic range enhancement. Figure 5-8 shows details of the electric guitar. The electric bass is similar, except for stringing.

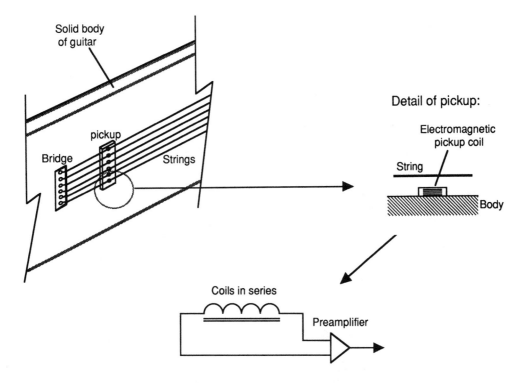

Figure 5-8. Details of the electric guitar.

In the recording studio, the engineer has several options for recording the various instruments fitted with pickups. The most popular method is to feed the output from the pickup directly into the recording console. This will give good isolation, insofar as there is no leakage into the input from other instruments. Another approach is to place a microphone close to the loudspeaker over which the pickup is amplified and presented. Problems here often include hum or buzz originating in the amplifier, or even distortion originating in the loudspeaker mechanism itself. Many amplifying systems for guitars include a certain amount of signal processing, and if it is desired to pick that up, then the only choice is to place a microphone in front of the loudspeaker. The traditional acoustical bass and guitar can be picked up in these manners as well, and it is common to use both direct and microphone feeds from these instruments so that the engineer and recording producer have several "ingredients" to work with in establishing a final sound and texture.

5.6 ACOUSTICAL CHARACTERISTICS OF STRING INSTRUMENTS

The data presented in this section represent a consensus of measurements that have been published over the years. Measurements are not always in agreement, and the author has taken the liberty of discarding or averaging data, as seems appropriate for each case.

5.6.1 Power Output

Table 5-1 shows the normal maximum acoustical power output of string instruments and the corresponding sound pressure levels referred to a distance of 1 m. Actual measurements for this table were made in the far field. To convert the level readings to any other distance, subtract the quantity 20 log r, where r is the distance in meters.

The dynamic range capability of the violin as a function of frequency is shown in figure 5-9a, again referred to a distance of 1 m. Note that the dynamic range of the instrument is about 30 dB and is fairly uniform over its frequency range. Similar data for the viola and cello are shown in figure 5-9b and c. The style of data presentation used here is based on that of Clark and Luce (1965) and Patterson (1974).

Table 5-1. Power and Sound Pressure in Outputs in Low and Middle Registers for Orchestral String Instruments

Instrument	Power, W	SPL (1 m)*
Violin	0.01	89
Viola	0.01	89
Cello	0.04	95
Bass viol	0.07	97.5

* Based on assumed directivity index of 0 dB (see section 1.10).
Source: Data after J. Meyer, *Acoustics and the Performance of Music*, translated by Bowsher and Westphal, Frankfurt: Verlag Das Musikinstrument, 1978.

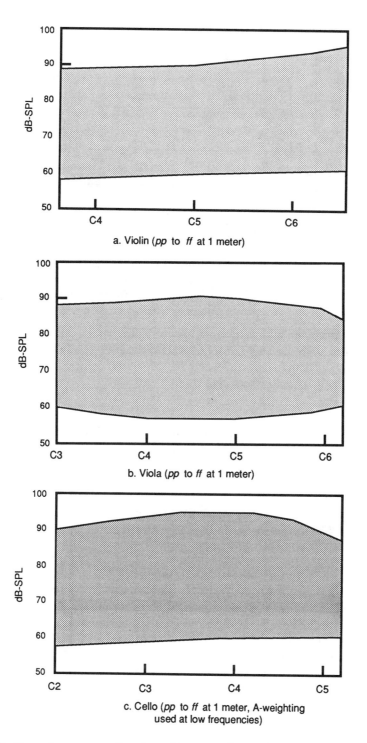

Figure 5-9. Dynamic range as a function of frequency for (*a*) the violin; (*b*) the viola; (*c*) the cello.

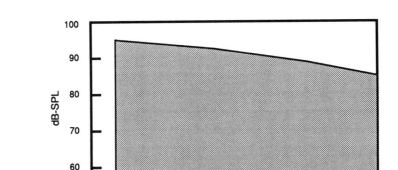

Figure 5-10. Dynamic range as a function of frequency for the bass viol.

Figure 5-10 shows similar data for the bass viol. Note that the output capability in the upper range falls off gradually from its lowest notes up to about C_4, due largely to the effects of string and body mass. The data in figures 5-9 and 5-10 are based on measurements by Meyer (1978), Clark and Luce (1965), and Patterson (1974).

5.6.2   Spectra and Directional Properties of String Instruments

The orchestral strings as a group produce a broad spectrum of sound whether they are played softly or loudly. In contrast, woodwind and brass instruments develop a broader spectrum the louder they are played.

The string spectrum varies considerably with frequency, and in most of the published data we can easily see an overall spectral envelope for each of the instruments. Figure 5-11 shows spectral data by Olson (1952) for the four open strings of the violin. The effect of the various body and Helmholtz resonances on the fundamentals and harmonics of the lower three strings can be easily seen, as can the extended high frequency output of the steel E string.

Figure 5-12 shows corresponding data for the bass viol. Note here that the overall spectral envelope of all the strings has a peak in the 100-Hz range. In recording, engineers often place a microphone fairly close to the bass viol in an effort to emphasize frequencies in the 40- to 80-Hz octave through microphone proximity effect (see section 13.2.3). The low frequency output of the instrument is highly dependent on nearby boundaries, increasing noticeably when the instrument is placed at the intersection of a wall and the floor.

In the case of directional measurements on string instruments, it must be stated at the outset that the fine structure of radiation patterns can, for a given instrument, change markedly over very small frequency and angular intervals. In these summaries, we present averages over one-octave frequency intervals and angular intervals of 15°. The intent is basically to show general trends.

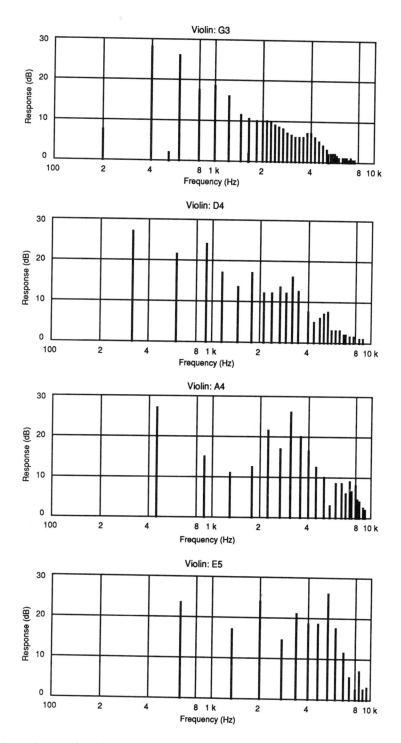

Figure 5-11. Spectra for the open strings of the violin.

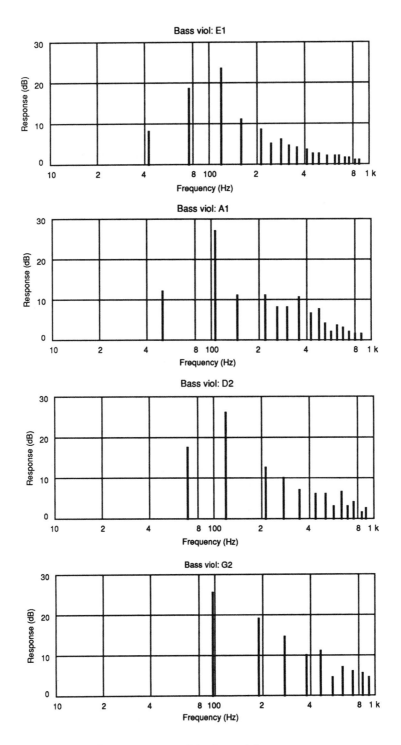

Figure 5-12. Spectra for the open strings of the bass viol.

Figure 5-13a shows the directional response of a violin in the plane of the bridge. The frequency ranges shown are on octaves centered at 500 Hz and 1, 2, and 4 kHz. The instrument is as seen from the position of the player. Horizontal directional data are shown in figure 5-13b.

Note that the instrument shows relatively little directionality in the 500-Hz range, while response in the 1- and 2-kHz octaves is maximum along the perpendicular to the top plate. At higher frequencies the directional response is characterized by many complex modes of vibration, and thus appears to be broad.

Most recording engineers avoid placing microphones along the axis of the instrument's top plate, as the sound may be too bright, or even irritating if the microphone is at close quarters.

The viola is similar to the violin in its directional properties.

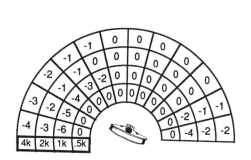

a

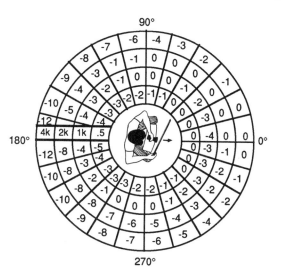

b

Figure 5-13. Directional characteristics of the violin, response normalized for each octave. (a) Vertical, in the plane of the bridge; (b) horizontal.

Figure 5-14. Directional characteristics of the cello in the horizontal plane; response normalized for each octave.

The directional response of the cello is shown in figure 5-14. There is preferential radiation perpendicular to the top plate in the 400- to 600-Hz range, corresponding generally to the violin's behavior in the 1- and 2-kHz ranges. The radiation of the cello is essentially spherical below about 200 Hz.

The directional characteristics of the bass viol are similar, but transposed downward according to the instrument's greater size. Its radiation pattern is spherical in the range below 100 Hz.

REFERENCES

Backus, J. 1969. *The Acoustical Foundations of Music.* New York: W. W. Norton.

Benade, A. 1976. *Fundamentals of Musical Acoustics.* New York: Oxford University Press.

Berg, R., and D. Stork. 1982. *The Physics of Sound.* Englewood Cliffs, NJ: Prentice-Hall.

Campbell, M., and C. Greated. 1987. *The Musician's Guide to Acoustics.* New York: Schirmer Books.

Clark, M., and D. Luce. 1965. "Intensities of Orchestral Instruments Played at Prescribed Dynamic Markings." *Journal of the Audio Engineering Society* 13(3).

Cremer, L. 1984. *The Physics of the Violin.* Translated by John S. Allen. Cambridge, MA: MIT Press.

Culver, C. 1956. *Musical Acoustics.* New York: McGraw-Hill.

Gill, D. 1984. *The Book of the Violin.* New York: Rizzoli.

Ginsburg, L. 1983. *History of the Violoncello.* Neptune City, NJ: Paganiniana Publications.

Hutchins, C. 1967. "Founding a Family of Fiddles." *Physics Today* 20.

Meyer, J. 1978. *Acoustics and the Performance of Music.* Translated by Bowsher and Westphal. Frankfurt: Verlag Das Musikinstrument.

Moravcsik, M. 1987. *Musical Sound.* New York: Paragon House Publishers.

Olson, H. 1952. *Musical Engineering.* New York: McGraw-Hill.

Patterson, B. 1974. "Musical Dynamics." *Scientific American* 231(5).

Peterlongo, P. 1979. *The Violin.* New York: Taplinger.

Pierce, J. 1983. *The Science of Musical Sound.* New York: W. H. Freeman.

Rossing, T. 1990. *The Science of Sound.* Reading, MA: Addison-Wesley.

ADDITIONAL RESOURCES

Lewin, F. "Music for the New Family of Violins," Musical Heritage Society MHS 4102.

The New Grove's Dictionary of Music and Musicians. London: Macmillan, 1980.

The New Harvard Dictionary of Music. Cambridge, MA: Harvard University Press, 1986.

The New Oxford Companion to Music. New York: Oxford University Press, 1983.

6

Acoustics of
Woodwind Instruments

The woodwinds include both reed and nonreed instruments, and the only common element is the use of keyed tone holes for changing pitch. At one time, they were all made of wood, but that has long since changed. Today, various metals are used, as are certain plastics. But out of this diversity comes a surprising blend of sound, and a fine woodwind section is the prize asset of a modern symphony orchestra.

These instruments have continued to be refined up to the present time, and the modern versions are capable of better intonation than their predecessors. As a group, they exhibit the narrowest dynamic range of the orchestra, a minor shortcoming that is more than made up for by their broad range of colors.

6.1 CLASSIFICATION OF WOODWIND INSTRUMENTS

While they are all aerophones, woodwind instruments can be classified according to bore profile and method of sound generation, as shown in Table 6-1.

6.2 TRANSPOSING INSTRUMENTS

During their early development, woodwind instruments were provided with rudimentary fingering systems, normally based on the diatonic scale of C major. With such a fingering system, the scale of C major could be played rapidly, raising or lowering one finger at a time. Playing in closely related keys was only slightly more complicated; however, playing in remote keys, such as B or F-sharp, was difficult, inasmuch as scale passages required simultaneous movement of at least two fingers for many adjacent notes.

Table 6-1. Classification of Woodwind
Instruments

	Bore Type	
	Cylindrical	Conical
Air reed	Flute	—
Single reed	Clarinet	Saxophone
Double reed	—	Oboe, English horn, bassoon

As a way around this problem, instrument makers scaled the instruments up or down in size, allowing the musicians greater ease in playing in remote keys. For example, an instrument pitched in A sounded a minor third lower than one pitched in C. Using the same fingering, and reading the notes of the C major scale, the player sounded the scale of A major. The player learned only one set of fingerings, and instruments so constructed were known as transposing instruments. The principle is shown in figure 6-1. The passage shown in figure 6-1a will sound as written on a nontransposing instrument. If it is played on a transposing instrument, say, in D, using the same fingering, and reading in the key of C, the passage will sound as shown in figure 6-1b. The term *concert pitch* is often used to indicate the actual note sounded by a transposing instrument. For instance, when an A clarinet plays the written note C_4, the note sounded is A_3 concert pitch.

Following is a list of the transposing woodwind instruments in use today. Those instruments that sound an octave higher or lower than the written part are not normally referred to as transposing. The piccolo, for example, sounds an octave higher than written, while the contrabassoon sounds an octave lower.

Instrument	Transposing interval relative to written note
Clarinet in E-flat	Minor third higher
Clarinet in B-flat	Major second lower
Soprano saxophone in B-flat	Major second lower
Clarinet in A	Minor third lower
Oboe d'amore	Minor third lower
Alto flute	Perfect fourth lower
English horn	Perfect fifth lower
Alto saxophone in E-flat	Major sixth lower
Tenor saxophone in B-flat	Major ninth lower
Bass clarinet in B-flat	Major ninth lower
Baritone saxophone in E-flat	Octave and major sixth lower

Figure 6-1. Principle of transposing instruments. The passage shown in (*a*) will sound as written on a nontransposing instrument. If the passage is fingered the same way on a transposing instrument in D, then it will sound as shown in (*b*).

Figure 6-2. Basic registers of the flute. Bracketed numbers refer to the overblown harmonic numbers.

6.3 THE FLUTE

The acoustical principle of the flute was explained in section 4.3. The instrument is held at a right angle to the mouth, and the player blows across the embouchure hole, forming a precise jet of air that impinges on the far edge of the hole. The hole is located about 25 mm (1 in) from a cork stopper at the end of the instrument. Early flutes were made of wood, but metals such as silver, gold, and platinum are favored today.

The instrument is about 66 cm (26 in) long and has a very flexible keying system devised in 1847 by the German instrument maker Theobald Boehm (1794–1881). The bore diameter of the instrument is about 19 mm, and the sound holes are large, requiring pads to cover them. The lowest note is normally C_4, but a B_3 extension [which increases the length by 5 cm (2 in)] is available.

The flute produces the complete family of harmonics, and it can be overblown as needed to the octave (second harmonic), twelfth (third harmonic), fifteenth (fourth harmonic), and seventeenth (fifth harmonic). There is no octave key to assist the player in reaching the upper registers. Rather, the player adjusts the embouchure by narrowing the jet of air. Precise control of the embouchure allows the player reasonable dynamic range in all registers. Figure 6-2 shows the specific registers of the instrument. Note that the overall range is just about three octaves, with D_7 as the practical upper limit.

The piccolo is pitched an octave higher, and its overall length is about 30.5 cm (12 in). Its range is from D_5 to D_8. The alto flute is just over 86 cm in length and has a range from G_3 to D_6. The rarely encountered bass flute is pitched an octave below the flute and has a length of 132 cm (52 in). Its range is from D_3 to G_5.

6.3.1   Performance Aspects

Instruments of the flute family are normally played with a vibrato. Rapid tonguing is possible, inasmuch as there is no mouthpiece to impede the movement of the tongue. Single tonguing (T-T-T, etc.), double tonguing (T-K-T-K, etc.), and triple tonguing (T-T-K-T-T-K, etc.) are all possible on the instrument. "Flutter" tonguing is an effect produced by a constant rolling of the tongue, as in the production of a rolled r sound in speech.

The term *multiphonics* refers to humming or growling at the same time a note is produced on the instrument. This produces an intermodulation between flute and voiced sounds and is useful in some contemporary writing. To a greater or lesser extent, the same technique can be used on all woodwind instruments.

Through careful embouchure control and partial closing of tone holes, some players are adept at producing slides between notes.

6.4   THE OBOE

The oboe has a double reed made of two pieces of thin cane placed back to back. Tone production is akin to pinching the end of a soda straw and blowing through it. Anyone who has ever done this can attest that there will be sound, but only with considerable air pressure, and there will not be much air velocity through the straw. The small opening of the reed is coupled to a conical resonator that is just under 61 cm (24 in) long. The tone holes are small, and the acoustical output of the instrument is not great. The instrument was the first woodwind to become a member of the orchestra, some 300 years ago.

The keying system is old and has never been successfully updated by the Boehm approach. The present fingering system is known as the "conservatoire" system because of developments made at the Paris Conservatory over the last 150 years.

The range of the oboe is from $A\#_3$ to G_6. Figure 6-3 shows the main registers of the instrument. There are two octave keys, operating automatically or semiautomatically, depending on the design of the instrument. The instrument is normally made of grenadilla or some other hard wood.

Since little air volume moves through the instrument, players must often stop and expel spent air from their lungs, take a new breath, and continue playing.

Other instruments in the family include the oboe d'amore and the English horn. Both of these instruments have a pear-shaped bell, which produces the formant for the sound *aw*,

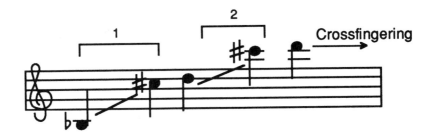

Figure 6-3. Basic registers of the oboe. Bracket 2 indicates the upper register, produced by octave keys.

giving these instruments their characteristic sound. The name English horn is a misnomer; the French term *cor anglais* is probably a corruption of "angled horn," referring to the bend in the metal crook connecting the reed with the body of the instrument.

The oboe d'amore has a length of about 63.5 cm (25 in) and a range from $G\#_3$ to E_6. The English horn has a length of about 81 cm (32 in) and a range from E_3 to A_5.

The oboe d'amore is widely used today in the performance of Baroque music. The bass oboe and its close relative, the heckelphone, are the largest members of the family, pitched an octave lower than the normal oboe. They are rarely encountered except in large scores by such composers as Richard Strauss (1864–1949) and Gustav Holst (1874–1934). The instruments of the oboe family are normally played with a vibrato.

6.5   THE CLARINET

The clarinet is the only modern wind instrument that behaves acoustically like the stopped pipe discussed in section 4.3.1. As indicated in the list in section 6.2, it comes in a variety of sizes; however, the models in A and B-flat are the primary ones. Both are used in modern scores, depending on the key of the music.

An excellent example of the use of transposing instruments is shown in figure 6-4. The passage shown in figure 6-4*a* is written in the key of C and is relatively easy to play. The first time the passage appears, it is in the key of A, and the A clarinet is used, producing the notes shown in figure 6-4*b*. The second appearance of the figure comes later in the score and is in the key of B-flat. Here, the player once again plays the passage in the key of C, but the B-flat instrument produces the notes shown in figure 6-4*c*. It would be virtually impossible to play the two passages on a single instrument at the normal performance tempo.

The B-flat instrument is a little over 67 cm (26 in) long. The bore is about 15 mm (0.6 in) wide and is basically cylindrical in profile. The bottom one-sixth of the bore flares out into a bell. The tone holes are fairly small, and several of the holes are directly covered by the fingers. The clarinet is normally made of African black wood, but lower-cost metal instruments are quite common in secondary schools.

The basic registers are shown in figure 6-5. Note that the instrument overblows to the twelfth (octave and a fifth), giving the lower register a wider range than any other single register in the woodwind family.

Figure 6-4. An interesting use of two transposing clarinets, A and B-flat. From Rimsky-Korsakov, *Capriccio Italienne.*

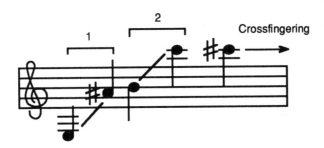

Figure 6-5. Basic registers of the clarinet. Bracket 2 indicates the upper register.

The clarinet embouchure calls for the upper teeth to be placed directly on the mouthpiece, while the lower lip is placed between the reed and the lower teeth. The player thus has considerable control of the pressure on the reed and its speaking length. Single tonguing is normal for the instrument, inasmuch as the presence of the mouthpiece in the mouth inhibits free motion of the tongue. Some players favor a double-lip embouchure, but this is rare.

The instrument is a relative latecomer to the orchestra, having found a permanent home there only during the time of Beethoven (1770–1827). Its tone color is unique, consisting predominantly of odd harmonics, and its dynamic range the greatest in the woodwind group.

The smaller E-flat instrument is capable of a particular "wailing" quality and has been employed as a coloristic element in program music where an "impish" quality is desired. The larger bass and contrabass instruments are S-shaped (like the saxophone) for ease in holding and playing.

6.5.1   Playing Characteristics

With its modern Boehm-derived fingering system, the clarinet can be played with great agility. In the symphony orchestra it is normally played without a vibrato, while in jazz or popular work a vibrato is common. Individual notes can be "bent" in pitch quite easily, and an effect known as the lip slur allows the player to produce a slide, or portamento, over a large range of notes. The effect is accomplished through flexible embouchure control as well as through the gradual uncovering of finger holes. The most famous example of this is at the opening of George Gershwin's (1898–1937) *Rhapsody in Blue*.

6.6   THE BASSOON

The bassoon is an old instrument, and its fingering system, like that of the oboe, has not been modernized. The most common form of the instrument is that of the German instrument maker Wilhelm Heckel (1856–1909), although a French form of the instrument is used in that country.

The bassoon is just about 245 cm (96 in) long and is doubled back on itself for ease in playing. The double reed is coupled to the instrument via a metal crook (bocal) that is quite narrow. The conical bore expands gradually through the four joints of the instrument, and the bore width at the bell is about 6 cm (2.3 in). The bassoon is supported by a sling worn around the player's neck.

Figure 6-6 shows a profile of the "wing" joint, in which close spacing of finger holes communicates with widely spaced internal openings along the bore. The lowest note of the instrument ($A\#_1$) is about a whole tone lower than would normally be produced by a conical resonator of the same physical length. Acoustically, the conical bore is rendered more complicated by the presence of many long passages leading to the finger holes, and this undoubtedly contributes to the difference between the physical and speaking lengths of the instrument. The bassoon is normally made of maple, but the French instrument is often made of rosewood.

In relation to the other double reeds, the main compass of the bassoon lies one octave lower than that of the English horn. This is the portion of the bassoon's range from F_2 to D_5. The extension from F_2 down to $A\#_1$ is controlled by keys operated mainly by the left thumb. These extended low notes were originally intended only for sustained bass lines, but in time composers began to write melodic passages in this lower range.

Figure 6-7 shows the main registers of the instrument. Three independent (nonautomatic) octave keys are necessary to achieve the full upper range of the instrument.

The contrabassoon is pitched an octave lower and is twice the length of the bassoon. Its keying system is complicated by the fact that the instrument bends back on itself four times.

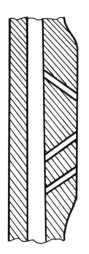

Figure 6-6. Profile of the bassoon wing joint.

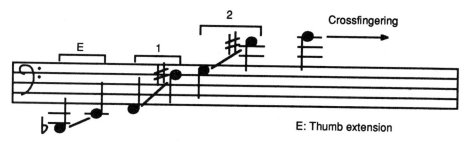

Figure 6-7. Registers of the bassoon. Bracket E indicates the bass extension; bracket 2 indicates the upper register, produced by octave keys.

6.7   THE SAXOPHONES

Invented by Adolphe Sax (1814–1894) in 1846, the saxophones were conceived as an integral family of woodwinds. They have found a lasting home in bands and in popular music. The alto and tenor models have been used as solo instruments in many classical orchestral works. They are not, however, considered permanent members of the symphony orchestra, inasmuch as they do not normally blend well with the traditional woodwinds.

The saxophones use a single reed, somewhat larger than that of a clarinet of similar range. The bore is conical and quite wide. The finger holes are all large and are covered by padded keys. As a result of its large dimensions, the instrument is capable of higher acoustical power output than any other woodwind instrument.

With its conical brass resonator, the instrument produces all harmonics. It overblows to the octave, and there are two automatic octave keys.

The familiar S-shape of the instrument is carried through all models except the small soprano saxophone, which is normally made straight. As can be seen from the list in section 6.2, there are many members of the family, all following the same basic plan of registers, as shown in figure 6-8.

As an acoustical result of its wide bore, the saxophone's pitch is not as tightly locked in to resonator length as those of the other woodwinds. As a result, the instrument can easily be altered in pitch through embouchure control, and this is an important element in jazz performance.

The instrument is usually played with a vibrato. In jazz and popular work the saxophone is capable of very high output levels, producing a timbre rich in harmonics that blends well with trumpets and trombones. In the hands of a master player, it is also capable of very expressive playing at quite low levels, producing a tone reminiscent of that of the English horn, or even the flute. Unfortunately, such playing is rare, and we would like to cite the superb work of Eugene Rousseau (see recording in Additional Resources).

The technique known as *circular breathing* is possible on a number of the woodwind instruments, and the author has observed it primarily with jazz saxophonists. In this technique, the player's cheeks are filled with air. That air is forced through the mouthpiece as the player breathes through the nose, refilling the lungs. Properly done, the transitions in air supply to the instrument are seamless, and tones can be sustained for many seconds. In the recording cited in Additional Resources, an alto saxophone player holds the note G_5 for approximately 45 seconds through circular breathing.

The photographs in figure 6-9 show various groups of woodwind instruments.

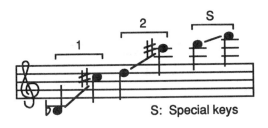

Figure 6-8. Registers of the saxophone. Bracket 2 indicates the upper register, produced by automatic register key; bracket S indicates the range produced by special keys.

a   b   c

Figure 6-9. Photographs of various woodwind instruments. (a) the piccolo; (b) the flute; (c) the oboe; (d) the clarinet; (e) the bassoon; (f) the saxophone. Photos courtesy of the Selmer Company.

d

e

f

6.8   ACOUSTICAL CHARACTERISTICS OF WOODWIND INSTRUMENTS

The data presented in this section are drawn largely from the work of Meyer (1978), Olson (1952), Patterson (1974), Culver (1956), and Clark and Luce (1965). Where appropriate, data have been averaged, and questionable data have been omitted.

6.8.1   Power Output

Table 6-2 shows the normal maximum power output of woodwind instruments and corresponding sound pressure levels referred to a distance of 1 m.

Figure 6-10 shows the overall dynamic range capabilities of the woodwinds, measured in sound pressure level referred to 1 m.

Note that the flute has a narrower dynamic range in the upper part of its frequency range than it does at lower frequencies. The reason for this is the necessity for overblowing to increasingly higher harmonics in order to achieve the higher registers. Such overblowing leaves little room for soft playing. The same situation exists with the piccolo (figure 6-10*b*), except that the overall dynamic range is even more limited.

The oboe (figure 6-10*c*) has a very narrow dynamic range at the bottom of its first register, expanding to a quite wide dynamic range toward the middle of the second register.

In the hands of an accomplished player, the clarinet has the widest dynamic range of any woodwind in the symphony orchestra, reaching about 50 dB in the lower part of the second register. Surprisingly, the data presented by both Patterson (1974) and Clark and Luce (1965) indicate that the clarinet has a relatively narrow dynamic range. We note this discrepancy; however, we show the data of Meyer (1978), inasmuch as his measurements were made with orchestral players of very high caliber.

While data are not available, it is felt that the alto saxophone would have dynamic range characteristics similar to those of the clarinet because of the similarities in embouchure control.

The bassoon has a generous dynamic range which, unlike those of the other members of the group, increases in the upper registers.

Table 6-2. Maximum Power and Sound Pressure Outputs of Orchestral Woodwind Instruments

	Power, W	Maximum SPL (1 m), dB
Piccolo	0.3	104
Flute	0.3	104
Oboe	0.1	99
Clarinet	1.0	109
Bassoon	0.2	102

Source: Data after J. Meyer, *Acoustics and the Performance of Music*, translated by Bowsher and Westphal, Frankfurt: Verlag Das Musikinstrument, 1978.

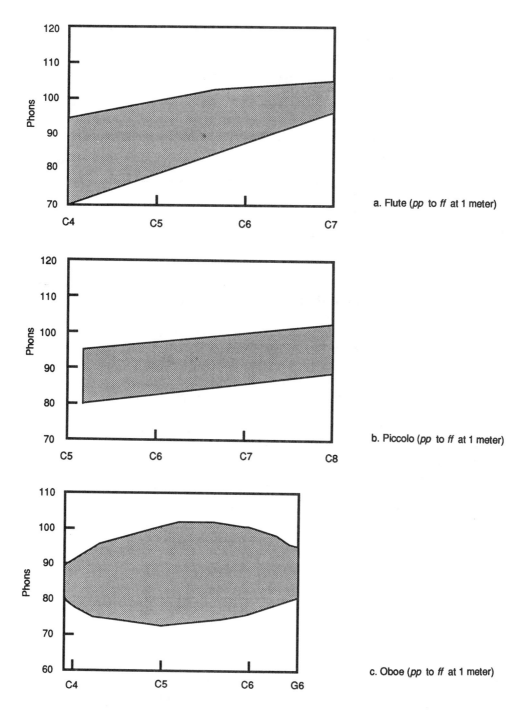

a. Flute (*pp* to *ff* at 1 meter)

b. Piccolo (*pp* to *ff* at 1 meter)

c. Oboe (*pp* to *ff* at 1 meter)

Figure 6-10. Dynamic ranges of woodwinds, all referred to a distance of 1 m. (*a*) Flute played from *pp* to *ff*; (*b*) piccolo played from *pp* to *ff*; (*c*) oboe played *pp* to *ff*; (*d*) clarinet played *pp* to *ff*; (*e*) bassoon played *pp* to *ff*, with C-weighting used for high frequencies and A-weighting used for low frequencies.

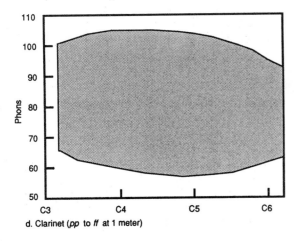

d. Clarinet (*pp* to *ff* at 1 meter)

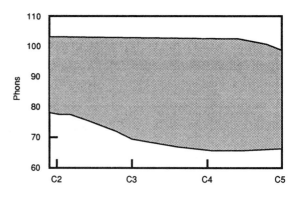

e. Bassoon (*pp* to *ff* at 1 meter; A-weighting used at low frequencies)

6.8.2   Spectra and Directional Properties of Woodwind Instruments

Figures 6-11 to 6-15 show the spectra for the various woodwind instruments. For the flute (figure 6-11), the fundamental is predominant in all registers shown. For the clarinet (figure 6-12), however, odd harmonics predominate in all registers, as does the fundamental.

For the oboe (figure 6-13), in the first two registers, the fundamental is not the strongest component. The component in the range of 1 kHz predominates in all three registers, taking on the properties of a formant frequency and giving the instrument its characteristic bright sound.

For the bassoon (figure 6-14), the fundamental is weak in the bottom two registers. This is due to the fact that the bore is quite small for its overall length and cannot support much fundamental acoustical output. Note further that there is a general prominence in the range from 400 to 800 Hz in all three spectra. Again, we have the effect of a formant, this time creating an *oh* sound. Meyer discusses this aspect of the instrument in detail (1978, 59–60).

For the alto saxophone (figure 6-15), the spectra are all rich in harmonics, indicating that the tones in these examples were all played at fairly high output. Only in the lowest register is the fundamental not the strongest component.

While sound radiation from the woodwinds is not as complex as that from the string instruments, it is made complicated by the presence of many tone holes, some of which may be open or closed, depending on the note played. Only the lowest note of a reed instrument produces radiation solely from the bell.

Benade (1985) presents the data shown in figure 6-16, in which sound radiation from a clarinet is divided into three ranges. At low frequencies the radiation is effectively omnidirectional. Above about 1500 Hz, the radiation is predominantly from the sides, while above 3500 Hz the radiation is along the axis of the bell. Benade states that the upper limit of omnidirectional radiation is controlled by a cutoff frequency, which is a function of bore diameter, tone hole spacing, and tone hole depth. The clarinet and oboe have similar cutoff frequencies around 1500 Hz, while the cutoff frequency of the alto clarinet and English horn is about 1 kHz. The bass clarinet cutoff frequency is about 750 Hz, and that of the bassoon is about 500 Hz.

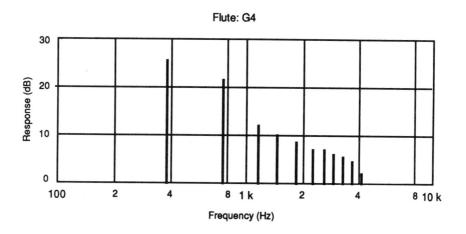

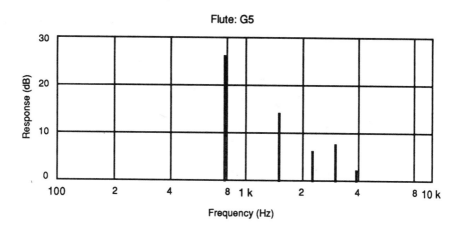

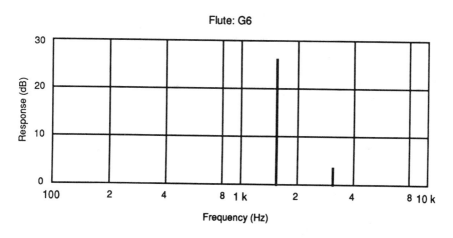

Figure 6-11. Spectra for the notes G_4, G_5, and G_6 as played on the flute. Data after Olson 1952.

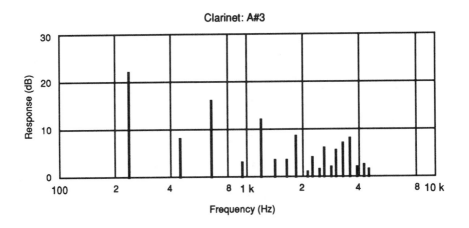

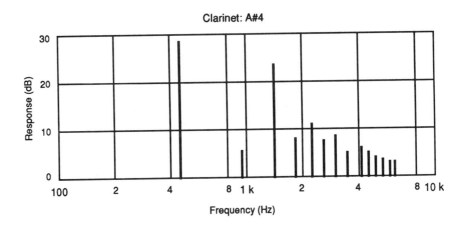

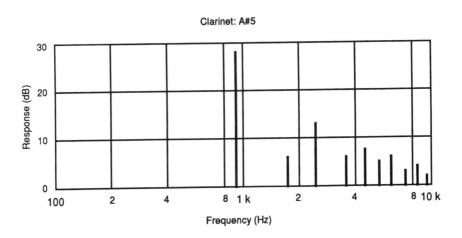

Figure 6-12. Spectra for the notes A#₃, A#₄, and A#₅ as played on the clarinet. Data after Olson 1952.

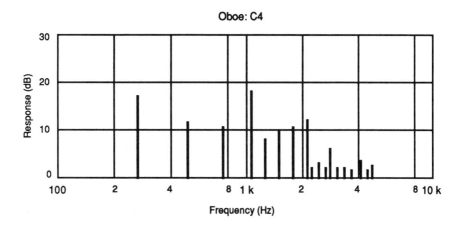

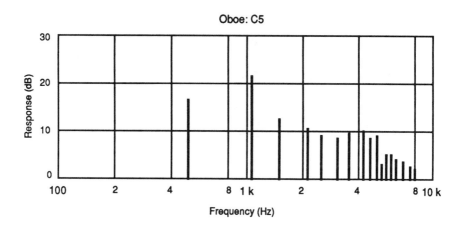

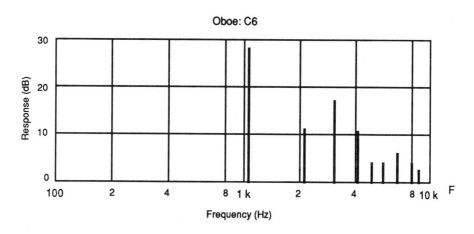

Figure 6-13. Spectra for the notes C_4, C_5, and C_6 as played on the oboe. Data after Olson 1952.

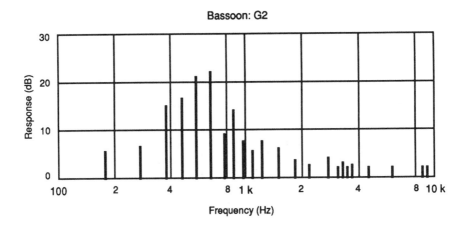

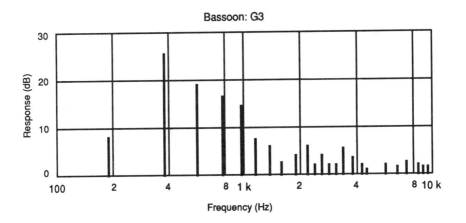

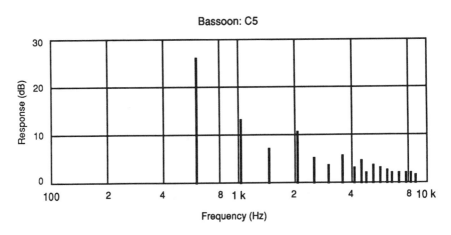

Figure 6-14. Spectra for the notes G_2, G_3, and C_5 as played on the bassoon. Data after Olson 1952.

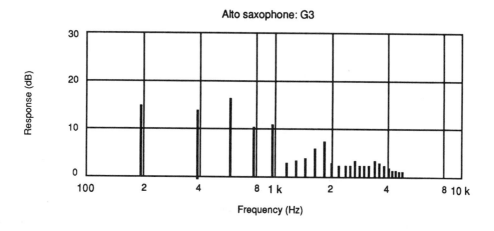

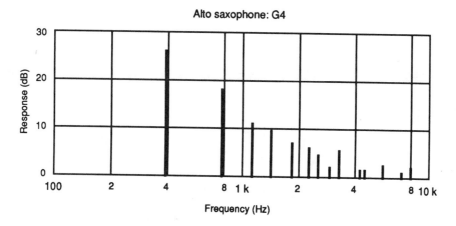

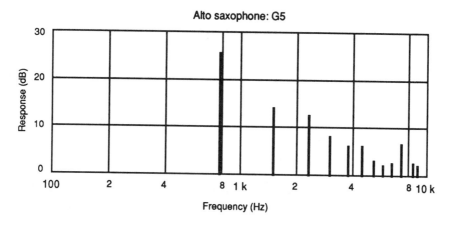

Figure 6-15. Spectra for the notes G_3, G_4, and G_5 as played on the alto saxophone. Data after Olson 1952.

It is surprising how many recording engineers routinely place microphones at the bell of woodwind instruments. It should be obvious that radiation is entirely from the bell only for the lowest note the instrument can produce! The technique works for the saxophone, however, since the position immediately above the upturned bell also places the microphone in a position to receive radiation from the tone holes. In many music reinforcement applications, microphones are attached to instruments so that the players will have freedom to move about the stage. Despite its drawbacks, the bell is just about the only place where a small microphone can be conveniently attached to the instrument.

Table 6-3 shows Benade's data in generalized polar form, as related to the cutoff frequency f_c. This information is valid for all reed instruments, taking into account the fact that the bells of some of these instruments are not coaxial with their bodies.

The flute is different from the other woodwinds in that sound radiation from the embouchure hole is about as strong as sound radiation from the region of the first open tone hole. The radiation from both effective sources is in phase for all harmonic components, and this indicates that radiation would be strongest in the direction perpendicular to the body of the instrument. At high frequencies (above about 6 kHz), the radiation predominates along the axis of the instrument.

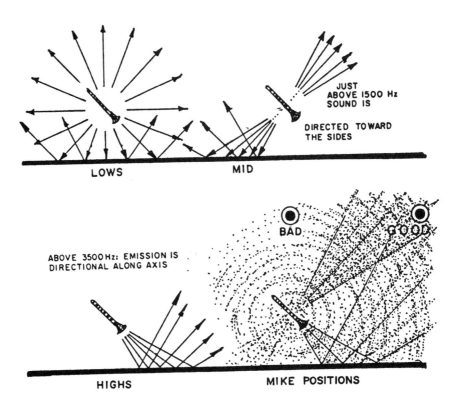

Figure 6-16. Directional characteristics of the clarinet and oboe. Data courtesy *Journal of the Audio Engineering Society* and Virginia Benade.

Table 6-3. Benade's Woodwind Directional Data in Polar Form

Angle θ	At f_c	At $4/3\,f_c$	At $2\,f_c$	At $4\,f_c$
0°	−40 dB	−7.4 dB	−2 dB	0 dB
±15°	−26	−5.6	−1	−1
±30°	−20	−2	0	−8
±45°	−6.5	0	−7	−18
±60°	−2	−2.5	−14	−18
±75°	0	−10	−16	−18
±90°	−2.5	−14	−20	−18
±105°	−8.5	−14	−26	−26
±120°	−14	−14	−26	−26

Note: Levels in each column are referred to 0 dB as maximum in each frequency; measurement distance is assumed to be valid for the far field.

As with all instruments, radiation from the woodwinds is influenced by nearby surfaces. In the case of the oboe and clarinet, whose bells normally face downward, the floor reflection is important at high frequencies. It is understandable why players generally dislike performing on carpeted surfaces.

REFERENCES

Backus, J. 1969. *The Acoustical Foundations of Music.* New York: W. W. Norton.

Benade, A. 1976. *Fundamentals of Musical Acoustics.* New York: Oxford University Press.

———. 1985. "From Instrument to Ear in a Room: Direct or Via Recording." J. *Audio Engineering Society* 33(4).

Berg, R., and D. Stork. 1982. *The Physics of Sound.* Englewood Cliffs, NJ: Prentice-Hall.

Campbell, M., and C. Greated. 1987. *The Musician's Guide to Acoustics.* New York: Schirmer Books.

Clark, M., and D. Luce. 1965. "Intensities of Orchestral Instrument Scales Played at Prescribed Markings." J. *Audio Engineering Society* 13(3).

Culver, C. 1956. *Musical Acoustics.* New York: McGraw-Hill.

Joppig, G. 1988. *The Oboe and the Bassoon.* Portland, OR: Amadeus Press.

Meyer, J. 1978. *Acoustics and the Performance of Music.* Translated by Bowsher and Westphal. Frankfurt: Verlag Das Musikinstrument.

Meylan, R. 1988. *The Flute.* Portland, OR: Amadeus Press.

Moravcsik, M. 1987. *Musical Sound.* New York: Paragon House Publishers.

Olson, H. 1952. *Musical Engineering.* New York: McGraw-Hill.

Patterson, B. 1974. "Musical Dynamics." *Scientific American* 231(5).

Pierce, J. 1983. *The Science of Musical Sound*. New York: W. H. Freeman.

Rossing, T. 1990. *The Science of Sound*. Reading, MA: Addison-Wesley.

ADDITIONAL RESOURCES

Brown, Ruth. "Have a Good Time," compact disc produced by Fantasy Records, FCD-9661-2 (reference: band 9, 6:37 to 7:22 minutes).

The New Grove's Dictionary of Music and Musicians. London: Macmillan, 1980.

The New Harvard Dictionary of Music. Cambridge, MA: Harvard University Press, 1986.

The New Oxford Companion to Music. New York: Oxford University Press, 1983.

Rousseau, Eugene. "Saxophone Colors," compact disc produced by Delos International, CD1007.

7

Acoustics of Brass Instruments

Brass instruments are in principle among the oldest aerophones known, the original ones quite literally having been hollowed out of animal horns. Because of their high acoustical output, these devices have been used for thousands of years for signal and ceremony in religious, civil, and military contexts. Military bugle calls remain with us today as a reminder of this past.

The first brass instruments to find their way into the orchestra were the natural horn and the trumpet. These instruments were played without valves or any other quick means of altering the tube length, and all notes were those that could be overblown high in the natural harmonic series of the instrument. As a rule, players had to adjust the pitch of some of these notes by "lipping" them up or down, as required. In the case of the natural horn, the right hand rested in the bell of the instrument and could be moved in or out to alter the pitch or quality of notes, as the player wished.

In time, valves and slides were incorporated into brass instruments, giving them the capability of playing relatively rapid scale passages. There were a few keyed brass instruments, such as the so-called serpent, and the keyed bugle. These were musically unsatisfactory and have no progeny today.

7.1 CLASSIFICATION OF BRASS INSTRUMENTS

Figure 7-1 shows the straightened-out profiles of representative brass instruments. The short portion of the bore to the left in the diagrams indicates the tubing between the mouthpiece and the section of the instrument that is variable in length, either through both valves and tuning slides or through a slide only (in the case of the trombone).

All the variable-length sections were cylindrical in profile, but the fixed sections of the bores varied from cylindrical to conical. The upper three profiles in the figure show current orchestral instruments, while the euphonium profile is characteristic of many large conical bore instruments common in band instrumentation. The orchestral tuba has a similar bore profile; if straightened out in the manner shown in figure 7-1, it would be about 4 to 6 m (13 to 19 ft) in length, depending on pitch.

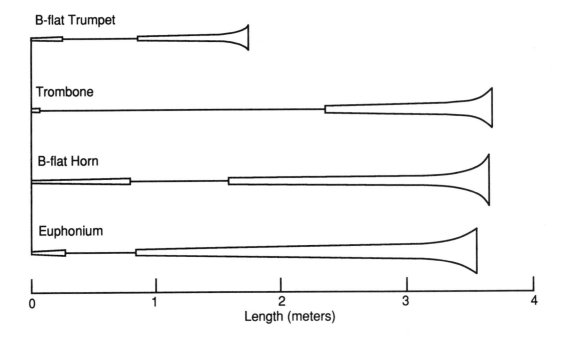

Figure 7-1. Bore profiles of major brass instruments.

Table 7-1. Categorization of Brass Instruments by Bore and Length

Approximate Length, m	Open Pitch	Cylindrical Bore	Horn Bore	Wide Conical Bore
1.2	B-flat	Trumpet	—	Fluegel horn
2.4	B-flat	Trombone	B-flat horn	Euphonium
3	F	Bass trombone	F horn	F tuba
4.8	B-flat	Contrabass trombone	—	B-flat tuba

Table 7-1 shows a further description of the brass family based on tube length (not including the variable section), open fundamental pitch, and general bore type.

Although all the instruments terminate in a conical section with a flaring bell, the main portion of the bore can be categorized into three types: cylindrical, horn type (narrow conical), and wide conical.

The cylindrical bore instruments in use today are the various trumpets and trombones. The effect of the cylindrical bore is a sound that is basically richer in harmonics than that produced with other profiles. The tone of these instruments becomes extremely rich in harmonics when they are played at high levels.

The horn bore starts with an extremely small aperture at the beginning of the lead pipe. This, in conjunction with the deep, tapered mouthpiece, gives the horn its characteristic sound.

The wide conical bore instruments are used mainly in bands. The tuba is the only representative of this group normally used in the orchestra.

The cornet and fluegel horn are related to the trumpet in that all three are of the same basic speaking length. The trumpet has the longest section of cylindrical tubing, and hence the brightest sound. The cornet is a close relative, and its longer conical lead-in section to the valves is sufficient to give the instrument a slightly more mellow sound. The fluegel horn has a very broad sound indeed, as a result of its wide conical bore.

As with the woodwinds, there are transposing brass instruments, as indicated below.

Instrument	Transposing interval relative to written note
Trumpet in E-flat	Minor third higher
Trumpet in B-flat	Major second lower
Fluegel horn in B-flat	Major second lower
Wagner tubas:	
Tenor in B-flat	Major second lower
Bass in F	Perfect fifth lower
Horn in F	Perfect fifth lower

7.2 CHANGING SPEAKING LENGTH: CROOKS, VALVES, AND SLIDES

The early natural instruments had crooks that could be inserted into the tubing to change the fundamental tuning of the instrument. However, in order to change the basic key of the instrument, the player had to take a certain amount of time out from playing to make the crook change.

Valves were introduced in the early 1800s as a quick method of changing from one crook to another. Today, there are basically two kinds of valve, piston and rotary; these are shown in figure 7-2a and b. The rotary valve provides generally smoother through-the-valve transitions than does the piston valve; it is used on the horn, where playing high in the harmonic series benefits from a path with fewer discontinuities. A third kind of valve, the so-called Vienna valve, is shown in figure 7-2c. Today, it is used only on horns used in Viennese orchestras. It is a cumbersome design, requiring two piston sections to insert one piece of additional tubing. Its virtue is that its "straight through" position offers no discontinuity in the tubing at all. Thus, when no valves are engaged, the Viennese horns perform essentially as natural horns.

Today, valves are normally used in a set of three. As seen from the player, the first valve lowers the pitch by one tone; the second valve lowers the pitch by a semitone, and the third valve lowers the pitch by three semitones.

A downward chromatic scale on the trumpet is fingered as shown in figure 7-2d. Note that the first five notes are produced by sounding the instrument's eighth harmonic, then adding

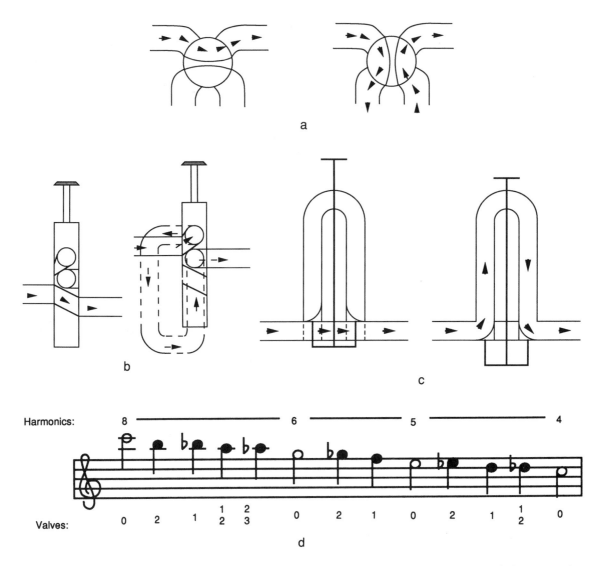

Figure 7-2. Valves for brass instruments: (a) Piston, (b) rotary, and (c) Vienna; (d) fingering of a descending chromatic scale on the trumpet.

the necessary tubing by means of the valves to lower the pitch chromatically. Then the sixth harmonic takes over for the next three notes. The fifth harmonic is used for the following four notes, and the process begins anew with the fourth harmonic. Obviously, there may be more than one way to finger a given note, and this is at the player's option.

The length of inserted tubing for each valve can be calculated exactly. For example, the second valve adds 5.9% to the speaking length of the tube in order to lower the pitch exactly one semitone. The first valve by itself adds 12.1% to the speaking length in order to lower the pitch one whole tone.

Now, if these two valves are engaged in order to reduce the pitch by three semitones, the increase in speaking length is the sum of the two valves, or 18%. But the proper value for this pitch decrease of three semitones is 18.8%, so there is a significant discrepancy. However, the demands of playing facility have won out, and most players will use valves one and two to produce a pitch decrease of three semitones. Therefore, the tuning slides on valves one and two have been adjusted in order to get a "best fit" for all notes. Valve number three has been calculated so that it adds the correct value of 18.8% to the speaking length, however, and is useful for alternate fingerings and facilitating trills.

When two or more valves are engaged, some adjustment in pitch will always be required. This can be done by embouchure control, or it may be done by the player's engaging small tuning slides with the left thumb and third finger, adjusting them as needed while playing. The main valves are operated by the first, second, and third fingers of the right hand on all valved instruments except the horn and the Wagner tubas.

The trombone makes use of a cylindrical slide for pitch adjustment, and exact pitches are easy to attain because of the continuous nature of the adjustment.

All brass instruments are capable of rapid tonguing of all types, including flutter tonguing, depending on the skill of the player.

As befits their name, all brass instruments are constructed of brass or related alloys, and are plated and lacquered for protection from corrosion.

7.3   THE HORN

The instrument we have referred to as the horn is often called the French horn, the national affiliation dating from its introduction into England from France in the late 1600s. In any musical context, however, it is always referred to simply as the horn.

The horn's shape is such that the bell bends back to form an angle with the mouthpiece and lead pipe of about 60°. Thus, the bell faces rearward to the player's right. The right hand is placed in the bell, and the valves are actuated with the left hand. The right hand got its position first, during the early development of the instrument.

As we have stated, the instrument plays high in its harmonic series. While the trumpet may play up to its eighth harmonic, the horn routinely plays from its fifth or sixth harmonic up to its sixteenth harmonic and beyond.

The mellow, full sound of the instrument is not always apparent if one stands behind the player. When heard in the concert hall, the sound is highly modified by the player's hand and by the fact that the sound reaches the hall entirely by way of reflected paths, which attenuate high frequencies, often making it difficult to localize the instrument precisely.

The instrument existed in its natural (valveless) form until well into the nineteenth century, and the scores of such composers as Johannes Brahms (1833–1897) abound in writing for the natural horn. In his scores we see indications for horns in D, horns in A, and so forth, indicating the crooks to be used to pitch the instrument in the desired key.

Thus, the tradition developed early of having two pairs of horns in the orchestra. The first and second players performed as a team, with both instruments crooked in the indicated key. While they were playing, the third and fourth horns were preparing their crooks for the next key change, and so forth. The tradition further developed that the first and third players would specialize in playing higher parts, while the second and fourth players took the lower parts.

This tradition holds today, and most horn players specialize in high or low playing, each requiring specific embouchure control.

The solution to the problem of multiple crooks came about when valves were added to the horn. The horn in F became one of the standards, and this has established the current transposing interval of the instrument. The F horn has an open length (without valves engaged) of about 3.75 m (about 12.3 ft). This model is still the one on which many students learn to play; however, in the orchestra, higher horn parts require another instrument. The horn in B-flat, pitched a fourth higher, is common here, and most of the horns of today are in effect double instruments, with two sets of valve tubing so that the player can pitch the instrument in either key simply by actuating a fourth valve.

From the point of view of part reading, however, the instrument today is still written as pitched in F, regardless of which set of tubing the player uses. It is the player's option to pick one set of tubing or the other, and horn playing requires a good bit of transposing at sight. Experienced horn players have become very adept at this.

The need for the two instruments can be seen from figure 7-3. Here, we show the actual sounding pitches as related to the harmonic series of both F and B-flat horns. Players are on dangerous ground when they rove too high in the harmonic series, and a passage that lies high on the F horn may be very comfortable on the B-flat instrument. For example, notes in the interval from G_4 to E_5 lie between the ninth and fifteenth harmonics on the F horn. By contrast, they lie between the sixth and eleventh harmonics on the B-flat instrument, and thus may be easier to play accurately on the shorter instrument. The length of tubing for the B-flat instrument is 2.8 m (9.2 ft).

For parts that lie extremely high, some players have become adept at playing a so-called descant F horn pitched an octave higher than the normal F horn. Again, regardless of which instrument is used, horn parts today are universally written as if the instrument sounded a perfect fifth below the written note. In other words, the indication "Horns in F" is always in effect.

Figure 7-4a shows the standard double horn in F and B-flat. Note the two sets of tubing, with the longer F tubing lying on the outside of the instrument. The effect of the F to B-flat change on routing through the instrument is shown in figure 7-4b and c.

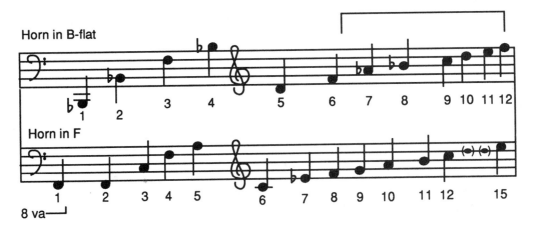

Figure 7-3. Useful ranges of the harmonic series on the F and B-flat horns.

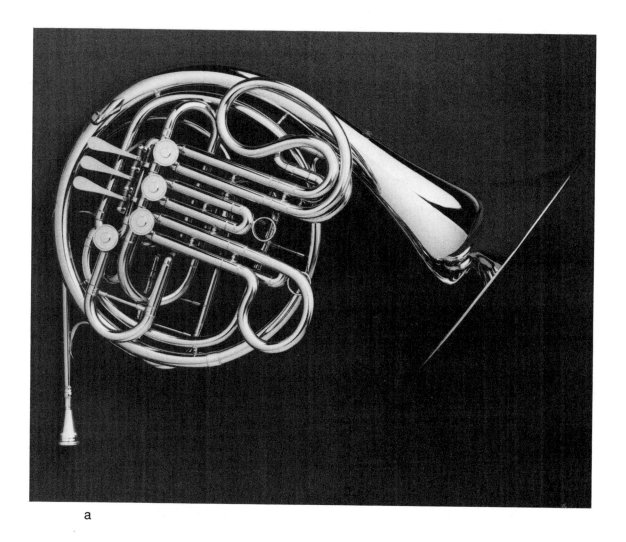

a

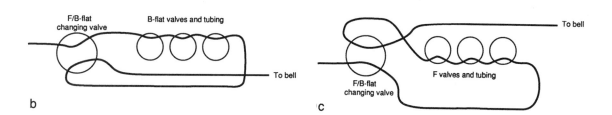

b

F/B-flat changing valve

B-flat valves and tubing

To bell

c

F/B-flat changing valve

F valves and tubing

To bell

Figure 7-4. The modern horn. (*a*) Double F–B-flat horn; (*b*) F horn tubing; (*c*) B-flat horn tubing. Photo courtesy of the Selmer Company.

7.3.1   Playing Technique

During normal playing of the instrument, the player will adjust the position of the right hand in and out of the bell as required to alter the pitch or quality of a tone. In orchestral parts there may be indications of muting the instrument with the hand, creating a quite forlorn sound. Further insertion of the hand, along with louder playing, will create a very nasal, metallic sound, unlike anything else the orchestra can produce.

Horn parts sometimes call for "bells in the air," which redirects the bells toward the overhead stage canopy, resulting in a brighter, often rough-edged sound.

Because the horn is played fairly high in its harmonic series, it is relatively easy to produce a lipped *glissando*, or glide, which produces a loud upward "swoop" of sound. Richard Strauss (1864–1949) and Igor Stravinsky (1882–1971) have used this to advantage.

The range of the horn varies widely with the player, but we can normally give that of the double horn as being from about F_2 to about F_5.

In recording horn concertos and other such pieces, care must be taken that the direct sound of the instrument not predominate. A microphone should never be placed behind the instrument, unless for very special effect. In general, a set of reflecting baffles should be located perhaps 2 m behind the player, in a normal forward-facing attitude, so that reflected sound from the instrument is dominant.

Certain chords can be played on the horn. If a player hums a note, say, a perfect fifth higher than the fundamental, two additional tones will be produced, a sum tone and a difference tone. When the player hums a perfect fifth above the horn tone, the frequency ratio will be 3 to 2. The sum tone will be 5, and the difference tone will be 1. Thus, the result will be a series of four tones with frequency ratios of 1, 2, 3, and 5, sounding a major chord. The effect is difficult to control and has rarely been used in composition for the instrument.

7.4   THE TRUMPET

The precursor of the modern trumpet was the Baroque trumpet, a natural instrument that, like the horn, played high in its harmonic series. The modern trumpet normally sounds notes between the third and eighth harmonics, using valves for chromatic intervals between the harmonics.

The B-flat instrument was standard for many years, but the nontransposing instrument in C has become popular in recent times. Many players prefer that instrument and routinely transpose B-flat parts for it, as required.

The trumpet produces a wide range of dynamics, and the change in timbre from soft to loud is quite striking. While it is often cast in a heroic role, it is capable of great finesse and expression, in both classical and jazz musical contexts. For fanfares and military calls, it is often located offstage, or antiphonally at some other position in the performance space.

Piston valves are normally used with the instrument, but in Germany rotary-valve trumpets are commonly encountered. The instrument is normally made of plated brass, and its overall length from mouthpiece to bell is about 46 cm (18 in.), depending on manufacture.

The smaller trumpets were developed in the nineteenth century to facilitate the playing of early music (mainly Handel and Bach) that called for very high trumpet parts. There were few Baroque trumpets available at the time, and even fewer players for them.

The bass trumpet is rarely encountered. There are several versions, one pitched an octave

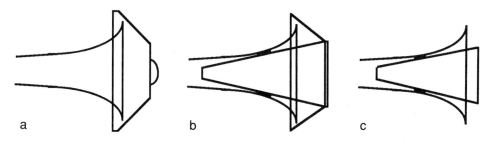

Figure 7-5. Various mutes for brass instruments: (*a*) Straight mute; (*b*) cup mute; (*c*) plunger.

lower than the standard C trumpet. As such, it has the overall tubing length of a trombone and requires a similar mouthpiece.

7.4.1   Mutes

While all the brass instruments may be muted by inserting the appropriate piece into the bell, it is the trumpet that has the greatest variety of mutes. As called for in classical writing, the so-called straight mute is normally employed. This is a conical plug made of fiber or plastic, with three or more small cork standoffs so that it can be inserted into the bell, wedged in place, and still allow a small passageway between the mute and the bell for sound to pass outward. The effect of the mute is to diminish the instrument's output in the low and middle ranges, while letting the very high frequencies speak out.

Other types of mutes alter only certain formant ranges, and some, such as the Harmon, or "wah-wah," mute, are variable with the player's left hand. On occasion, a cup-shaped device, even a hat, may simply be held in front of the bell for a mild degree of muting. Exotic mutes are largely confined to jazz performance. Figure 7-5 shows representative kinds of mutes. The straight mute (figure 7-5*a*) is normally used in orchestral writing, while the cup mute (figure 7-5*b*) and plunger (figure 7-5*c*) are used in jazz performance. This last mute is held over the bell by the player's left hand. The bell may be covered or uncovered, as required by the music.

7.5   THE TROMBONE

The trombone is a descendant of the *sackbut*, an instrument of slightly smaller scale. It was the first brass instrument to have a full chromatic compass. The slide is operated by the player's right hand, and downward chromatic steps are produced by adjusting the slide to successive positions. First position represents the unextended slide, second position adds enough tubing to lower the pitch one semitone, third position lowers the tone an additional semitone, and so forth. Although it is a nontransposing instrument, the standard tenor trombone is pitched in B-flat; that is, first position produces the harmonic series based on B-flat (58.2 Hz).

Many modern instruments have one or two extra sets of tubing that can be inserted into the instrument by valves operated by the left thumb. These effectively convert the instrument into another key for first position. Quite common here is the F conversion, lowering the instrument by a perfect fourth. Of course, the distance between successive positions of the slide will change considerably when this is done, since the basic length of the tube has been increased. Players become adept at this, knowing just the right "feel" for the various positions of the slide.

The bass trombone is essentially the tenor trombone in somewhat larger scale with extra tubing for converting it to F or to E-flat. Many varieties are encountered.

In jazz work, the valve trombone has attained some popularity because it allows the player the same kind of agility that we normally associate with the trumpet. The sound is not quite as smooth and rich as that of the traditional instrument. Mutes for the trombone are essentially scaled upward in size from those used with the trumpet.

The range of the tenor trombone is normally E_2 to $A\#_5$, while that of the bass trombone is C_2 to $A\#_5$.

7.6   THE TUBA

There are several models of tubas, all covering the same basic frequency range. They differ in scaling, and hence in sound quality. The instrument is held in the player's lap, and the bell points upward.

The tuba has a wide conical bore and therefore produces a tone with more fundamental than that of any of the other orchestral brass instruments. It produces a fitting bass line for the brighter upper instruments, because it has few harmonics to interfere with the musical and chordal textures of the upper instruments. It is capable of considerable volume, but with some roughness in the sound.

The standard instrument, if there is one, is the double B-flat model, with a tube length of about 5.5 m (18 ft). The instrument normally has four valves, with the fourth valve lowering the pitch of the instrument by a perfect fourth. The normal range of the instrument is from D_1 to F_4.

While the instrument may appear large and unwieldy, it is capable of rapid articulation at low and moderate outputs.

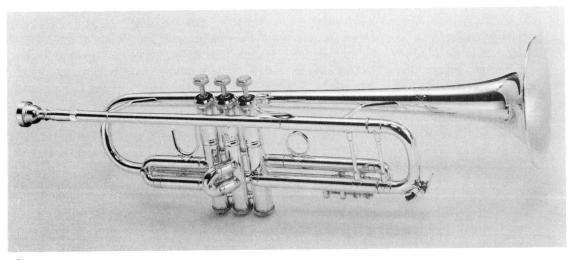

a

Figure 7-6. Photographs of various brass instruments. (*a*) The trumpet; (*b*) the trombone; (*c*) the tuba. Photos courtesy of the Selmer Company.

b

c

The so-called Wagner tubas are hybrid instruments that use the traditional horn mouthpiece. They begin with a small aperture and end with a bell of moderate flare. The valves are operated by the left hand, since the instruments were intended to be doubled by horn players five through eight in the large Wagnerian orchestra. The instruments do not play high in their harmonic series, as does the horn. The tenor form of the instrument has the range of the euphonium, which is a popular band instrument, while the bass form of the instrument is pitched as the F horn. The instruments have found little application outside the works of Richard Wagner (1813–1883), Anton Bruckner (1824–1896), and Richard Strauss (1864–1949).

Figure 7-6 shows photographs of the various groups of brass instruments.

7.7   ACOUSTICAL CHARACTERISTICS OF BRASS INSTRUMENTS

Data in this section are drawn from the work of Meyer (1978), Olson (1952), Patterson (1974), Culver (1956), Benade (1985), and Clark and Luce (1965). In some cases, data have been averaged.

7.7.1   Power Output

Table 7.2 shows the maximum power outputs and corresponding levels referred to a distance of 1 m. (The directivity index is assumed to be 0 dB.)

Figure 7-7 shows the overall dynamic range capabilities of the orchestral brass instruments, measured in sound pressure level referred to 1 watt.

As we have seen with the woodwinds, the output levels in those registers that require overblowing are higher than those in the lower ranges. Furthermore, the dynamic range tends to be slightly less in the upper ranges.

From low to high, the horn has the greatest dynamic range of the group; however, within a given frequency region, the dynamic range of the horn is limited to about 33 dB.

7.7.2   Spectra and Directional Properties of the Brass Instruments

Figures 7-8 to 7-11 show spectra for the various brass instruments. For the trumpet (figure 7-8), note that in the lowest note of the set, the fundamental is about 10 dB below the second

Table 7-2. Maximum Power and Sound Pressure Output Levels of Brass Instruments

	Power, W	Maximum SPL (1 m), dB
Trumpet	2.5	116
Horn	0.2	105
Trombone	5	119
Tuba	1	112

Source: Data after J. Meyer, *Acoustics and the Performance of Music,* translated by Bowsher and Westphal, Frankfurt: Verlag Das Musikinstrument, 1978.

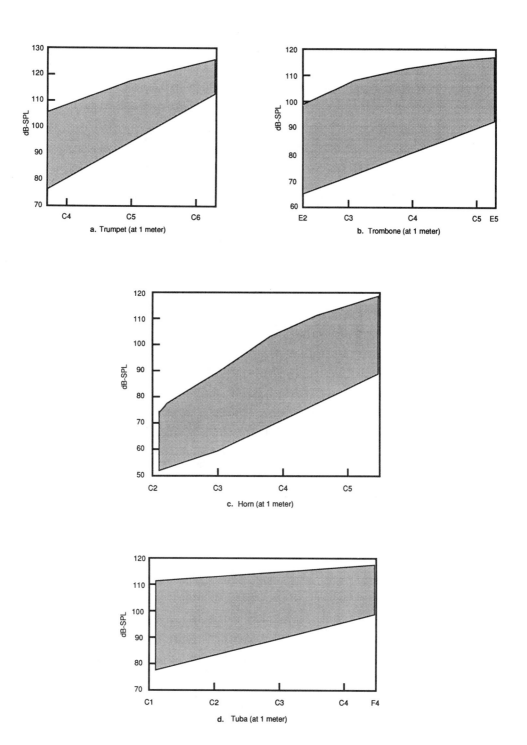

Figure 7-7. Dynamic ranges of orchestral brass instruments: (*a*) Trumpet; (*b*) trombone; (*c*) horn; (*d*) tuba.

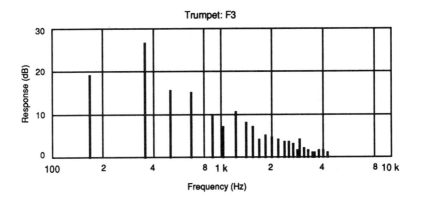

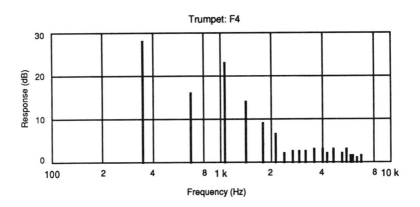

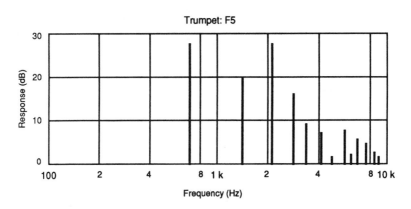

Figure 7-8. Spectra for the notes F_3, F_4, and F_5 as played on the trumpet. Data after Olson 1952.

harmonic. In the upper tones, there are strong formants in the 1- and 2-kHz range, giving the instrument its characteristic bright sound.

The fundamental and second harmonic of the trombone (figure 7-9) are of equal value at this frequency. At higher frequencies, this would also be the case, while at lower frequencies, the second harmonic would overshadow the fundamental.

For the horn (figure 7-10), the fundamental is about 10 dB below the second harmonic and the second harmonic predominates.

The major fundamental output of the tuba (figure 7-11) is in the range of about 100 to 300 Hz. The prominence in the 200- to 250-Hz range gives it a formant characteristic of the vowel sound *oo* (as in the word *cool*).

We remarked earlier that the spectrum of the cylindrical-bore brass instruments in particular develop higher harmonics the louder the instrument is played. We show this in figure 7-12, in which data from Benade (1985) have been replotted in decibels. The note C_4 is played at a number of loudness levels, from *pp* to *fff*. Note that the prominence at high playing levels reaches a maximum in the range of the seventh harmonic, which corresponds to about 1800 Hz. We also noted this tendency in the spectra shown in figure 7-8.

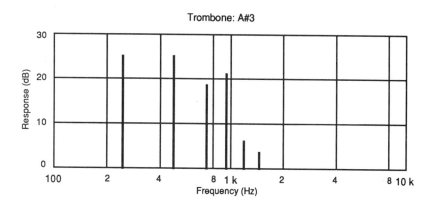

Figure 7-9. Spectrum for the note $A\#_3$ as played on the trombone. Data after Culver 1956.

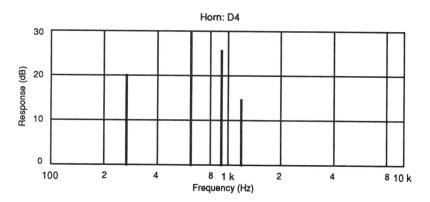

Figure 7-10. Spectrum for the note D_4 as played on the horn. Data after Culver 1956.

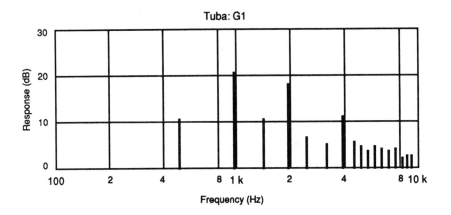

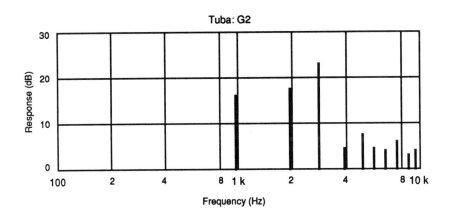

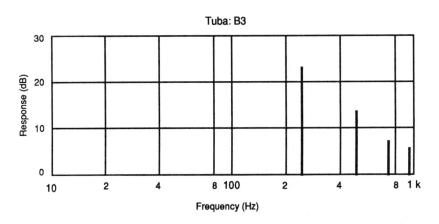

Figure 7-11. Spectra for the notes G_1, G_2, and B_3 as played on the tuba. Data after Olson 1952.

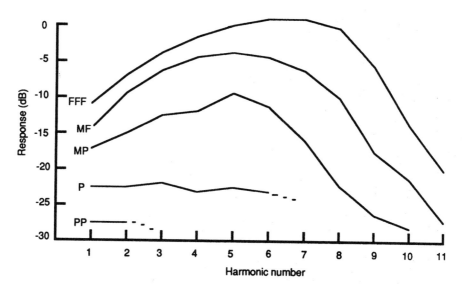

Figure 7-12. Spectrum versus playing level for the trumpet. Data after Benade 1985.

Since virtually all sound radiation from brass instruments is by way of the bell, we can easily relate the directional properties of the instruments by calculating bell diameter and wavelength relationships.

Benade (1985) presents directional data, which we have transcribed into the polar form shown in Figure 7-13. The data are based on a reference frequency f and are calculated for f, $2f$, $4f$, and $8f$, all representing octave increments above f.

For the trumpet, f corresponds to 500 Hz. For the trombone, f corresponds to 250 Hz, and for the tuba, f corresponds to 167 Hz.

These data will facilitate calculation of the directional properties along the axis of the bell of a brass instrument. However, we must be aware that local obstacles, floor reflections, and the like will modify the directional characteristics as observed at a distance. In particular, the presence of the player's hand in the bell of the horn means that the data are useless for that instrument.

Meyer (1978) has pointed out that at the highest harmonics, the directivity of the trumpet is so pronounced along its major axis that its critical distance is generally greater than the distance from the stage to the rear of the concert hall! What this means in practical terms is that a listener in the back row can detect by ear the movement of the bell as the player points it toward or away from him.

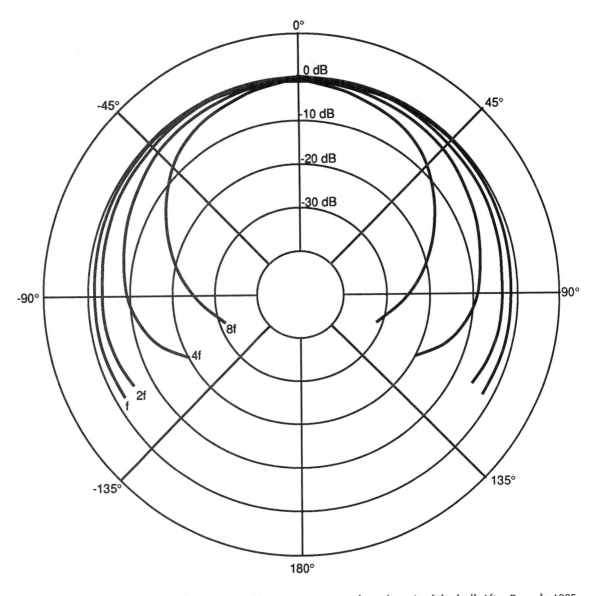

Figure 7-13. Directional properties of brass instruments along the axis of the bell. After Benade 1985.

REFERENCES

Backus, J. 1969. *The Acoustical Foundations of Music*. New York: W. W. Norton.

Benade, A. 1985. "From Instrument to Ear in a Room; Direct or Via Recording," *J. Audio Engineering Society* 33(4).

————. 1976. *Fundamentals of Musical Acoustics*. New York: Oxford University Press.

Berg, R., and D. Stork. 1982. *The Physics of Sound*. Englewood Cliffs, NJ: Prentice-Hall.

Campbell, M., and C. Greated. 1987. *The Musician's Guide to Acoustics*. New York: Schirmer Books.

Clark, M., and D. Luce. 1965. "Intensities of Orchestral Instrument Scales Played at Prescribed Markings." *J. Audio Engineering Society* 13(3).

Culver, C. 1956. *Musical Acoustics*. New York: McGraw-Hill.

Gregory, R. 1961. *The Horn*. London: Faber & Faber.

Janetzky, J., and B. Bruechle. 1988. *The Horn*. Portland, OR: Amadeus Press.

Meyer, J. 1978. *Acoustics and the Performance of Music*. Translated by Bowsher and Westphal. Frankfurt: Verlag Das Musikinstrument.

Moravcsik, M. 1987. *Musical Sound*. New York: Paragon House Publishers.

Morley-Pegge, R. 1960. *The French Horn*. London: Ernest Benn, Ltd.

Patterson, B. 1974. "Musical Dynamics." *Scientific American* 231(5).

Olson, H. 1952. *Musical Engineering*. New York: McGraw-Hill.

Pierce, J. 1983. *The Science of Musical Sound*. New York: W. H. Freeman.

Rossing, T. 1990. *The Science of Sound*. Reading, MA: Addison-Wesley.

Tarr, E. 1988. *The Trumpet*. Portland, OR: Amadeus Press.

ADDITIONAL RESOURCES

The New Grove's Dictionary of Music and Musicians. London: Macmillan, 1980.

The New Harvard Dictionary of Music. Cambridge, MA: Harvard University Press, 1986.

The New Oxford Companion to Music. New York: Oxford University Press, 1983.

8

Acoustics of Percussion Instruments

Percussion instruments have been with man from his earliest days and were central to religious activities and communications. Almost anything that made a sound when struck could be termed a percussion instrument, from a hollowed-out tree trunk to a suspended piece of rock. In those days, no special talents were required, and anybody could make an acceptable sound. Today, percussion instruments have taken on their own complexities and require skills equal to those of players in other instrumental groups.

Modern percussion instruments cover a very wide range in all regards, but those played by keyboards are generally relegated to a category of their own. In this chapter we will discuss the major percussion categories in terms of construction and playing methods. We will also describe the acoustical output characteristics of these instruments.

8.1 CLASSIFICATION OF PERCUSSION INSTRUMENTS

We can classify the instruments by using Sachs' definitions of membranophones and idiophones. Idiophones can be further broken down into metallophones and xylophones, depending on whether the vibrating structure is metal or wood (or other organic materials). All categories can further be classed as having either definite or indefinite pitch.

Table 8-1 shows the common instruments that fit within Sachs' framework. The listing is not exhaustive, and many folk instruments are omitted, since they are not normally encountered.

We will begin our discussion with the various membranophones.

8.2 MEMBRANOPHONES: THE DRUM FAMILY

The earliest member of this group to enter the orchestra was the timpani. The term refers to a set of tunable drums (often called kettledrums) that have the membrane (head) stretched over a metal hemispherical shell with a small vent hole in the bottom. An individual drum is tunable over an interval of about a perfect fifth by tightening or loosening the head with a series of adjustments about its periphery. The head has traditionally been made of calfskin, but many players today use plastic materials.

In the early days the pitch adjustment took some time to make, and the pitch of a given drum was not normally changed within a movement of a musical work. During the classical period, two drums were usually employed, one tuned to the tonic and the other to the dominant of the scale.

In time, more drums were added to the set, and quicker pitch changes were required. Today, pedal timpani are used, providing rapid and continuous pitch change simply by the player's operating a pedal that increases or decreases tension in the drum head.

Timpani are unique among the membranophones in that they produce sounds of quite definite pitch. The other members of the group produce sounds of indefinite pitch, although they may easily be judged as "higher" or "lower" relative to one another.

As we discussed in section 4.6, the normal modes of a circular membrane are not harmonically related, and the ear cannot readily assign a definite pitch to the sound. However, when the membrane is virtually sealed on the back side, as in the case of the timpani, the modes are redistributed in such a way that certain lower ones take on a harmonic relationship, making pitch identification quite easy. Since the enclosed air makes it difficult for the head to move in and out as a whole, all modes that would normally have such a component of motion are effectively blocked, leaving only those modes that result in the internal air mass moving with side to side motion. By this action, mode structure is significantly altered.

The lowest effective mode (the tuning note) is the one in which the enclosed air makes a simple back-and-forth motion within the hemisphere. Another mode exists that is 1.5 times the lowest, and there is still another mode that is about 1.9 times the lowest. These produce a set of frequencies roughly in the ratio of 2, 3, and 4, and the ear can easily lock in on the fundamental pitch. Other nonharmonic tones are present as well, giving the drum its distinctive sound.

The precise quality of a given drum will vary somewhat with its tuning, inasmuch as the ratios between modes will change slightly with retuning. Figure 8-1a shows a set of timpani, with corresponding frequency ranges shown in figure 8-1b.

Table 8-1. Classification of Percussion Instruments

	Unpitched	Pitched
Membranophones	Bass drum	Timpani
	Snare drum	
Metallophones	Cymbals	Gongs
	Tam-tam	Orchestra
	Triangle	bells
		Chimes
Xylophones	Wood block	Xylophone
		Marimba

a

b   Small Medium Large

Figure 8-1. (*a*) The timpani set; (*b*) typical tuning ranges. Photo courtesy of the Selmer Company.

The timpani are normally played with a pair of sticks with felted ends. Various hardnesses are available, depending on musical requirements. Single alternating left-right strokes are normal, and there is much crossing of hands. The head is struck about one-fifth to one-quarter of the way in from the rim, and the specific striking point has a profound effect on the quality of the tone. The pedal drum is capable of glissando effects, and these have been used to advantage by many modern composers, specifically Béla Bartók (1881–1945). Parts for the timpani are written in the bass clef at the sounding pitch of the drums.

The orchestral bass drum is quite large, often about 120 cm (47 in) in diameter and 45 cm (18 in) across. There are heads on both sides, and mode structure is a complex interaction between the two heads and the enclosed air mass. Fundamentals can extend down to the 30- to 40-Hz range, providing a substantial foundation for loud orchestral passages. A single-headed variation, known as the gong drum, is used by some European and English orchestras and is often used in recording. The gong drum can produce a low fundamental, but its decay is quite rapid.

The normal method of playing the bass drum is with single strokes of one hand. Rolls require two sticks, with both hands striking alternately. A well executed bass drum roll provides a continuous low-frequency underpinning for the orchestra, not unlike a low organ note.

The snare drum (or side drum) may be no more than 25 to 30 cm (10 to 12 in) in diameter and 12 to 15 cm (5 to 6 in) in depth. The head on the bottom has a ribbon of gut or metal strings, called snares, that can be tightened hard against the bottom head. When this is done, a stroke on the upper head produces not only motion of the heads but also a rattling of the snares against the bottom head. Snare tension can be adjusted for various effects, or the snares can be released completely.

The drum is played with a pair of wood sticks designed for good balance in the player's hands. Because of the resilience of the upper head, the sticks bounce quite readily; this action is used to advantage in a number of the basic playing rudiments. The familiar snare drum roll is produced by successive bounces of the sticks between the hands: LLRRLLRR, and so forth.

The field drum is basically a heavier drum of about the same diameter, but about 40 to 45 cm (16 to 18 in) in depth, producing a sound louder and deeper in pitch. It is the familiar drum of marching bands. Snares may or may not be used.

Other membranophones encountered in the orchestra are the tom-toms, bongo drums, and tambourine. Tom-toms have double or single heads and come in a variety of sizes, but are smaller than the field drum. The single-headed tom-tom has a rather precise pitch, and tuned sets have been made.

The bongo drum is clearly a Latin instrument and has found its way into the orchestra in music with an ethnic flavor. The single head is about 15 to 20 cm (6 to 8 in) in diameter, and the shell of the instrument is quite deep. It is played entirely by finger and hand contact.

The tambourine is a shallow wooden hoop with a single head and sets of metal "jingles" inserted several places in the hoop. It is played by shaking the instrument with one hand and tapping the head with the other. There are all sizes, but the normal orchestral version is about 40 cm (16 in) in diameter. The instrument has ancient roots in Mediterranean lands, but its role in the orchestra has gone far beyond the ethnic. Steady shaking of the instrument produces a sustained roll of the jingles. Running the moistened thumb across the head in a circular motion produces a hissing and rattling sound, not unlike that of a snake!

8.3 METALLOPHONES

The pitched metallophones in the orchestra include the tubular bells, the orchestra bells (glockenspiel), and the vibraphone. On special occasions, actual cast bells may be called for.

The instruments are laid out in keyboard fashion, chromatically arrayed from bottom at the left to top at the right. They are almost always tuned in equal temperament and cannot be changed. The tubular bells hang vertically, suspended on strings, and are struck with a hammer. The sound is more chimelike than bell-like, lacking the deep resonance of low bells of the same pitch. A damping mechanism can be used to prevent excessive ringout of the sound.

The orchestra bells consist of rectangular metal bars without resonators. The basic vibrational modes are as given in section 4.7. Generally, they are played softly with mallets of moderate hardness, producing a tone consisting mainly of the fundamental. If they are played loudly, or with a very hard mallet, some of the higher, nonharmonic modes will be produced. The higher modes generally die out much more quickly than the fundamental.

The vibraphone consists of large metal bars, each with a tuned cylindrical resonator below. At the top of each resonator is a small circular piece mounted on a thin shaft that runs the entire length of the instrument. There are two such assemblies, one for the diatonic scale of C and the other for the remaining notes in the chromatic scale. Both shafts can be motor driven at various speeds, thus providing a cyclic blocking and opening of the resonators beneath each bar. The sound thus takes on a pleasant amplitude modulation. For many applications, the rotors are turned off. The vibraphone is a staple in jazz, but its use in the symphony orchestra has been limited. Different mallet hardnesses provide a variety of sounds. A damper pedal controls the decay of the bars.

Bells of characteristic shape are normally associated with fixed installations in the steeples and towers of churches and civic structures. Bell modes are complex, and the sound is rich in nonharmonic overtones, making for glorious cacophony! Various materials have been used in their construction, but bronze is typical.

For pealing, bells are normally turned about a horizontal axis with a rope, creating an almost random effect, complete with Doppler frequency shift if the bell turns fast enough. For music performance, a set of bells known as a carillon is used. These are a chromatic set of fixed bells spanning two or three octaves, located in a suitable tower, played by clappers mechanically connected to a nearby keyboard. Such bells have a mode structure that allows them to sound together, preserving musical harmonic relationships.

The nonpitched metallophones include the following instruments.

The triangle is a steel bar [length about 25 cm (1 in)] that has been bent in the form of a triangle. The ends do not join, however. It is struck with a metal rod and produces a series of modes that are extremely high-pitched. Even at moderate playing level, it can be heard well over an orchestral tutti.

The cymbals are slightly concave bronze plates that are normally struck together with a sweeping motion. This is the so-called cymbal crash that is a part of so many orchestral climaxes. Other methods of playing include suspending the cymbal and rolling with soft sticks. Another effect is achieved by striking the suspended cymbal with a triangle beater and damping the sound immediately with the thumb and forefinger. Cymbals come in many sizes, the average diameter being about 40 to 50 cm (16 to 20 in).

The tam-tam and gong are clearly of oriental origin, but have been widely used in music of nonethnic connotation. The tam-tam is a large, slightly curved plate bent back on itself at the edges and suspended at two places on the edge. When struck moderately with a stick of medium hardness, it produces a rich, rolling sound, which continues to develop its mode structure for a few seconds after the stroke. In musical context, the tam-tam often conveys a sense of abject finality, and as such is used sparingly.

Gongs often come in a variety of sizes and pitches. They are generally smaller than tam-tams and have a slightly raised portion in the center. They can be of definite pitch and may be used in tuned sets.

8.4   XYLOPHONES

While there is an instrument called the xylophone, the term, as used by Sachs, describes any kind of "sounding wood."

The pitched xylophones include the marimba and the xylophone itself. The marimba has large wooden bars mounted above tuned resonators. Since wood is inherently more damped than metal, the tone produced by a single stroke dies out rather quickly. For a sustained effect, the marimba is generally played by rolling individual notes or chords. It is not uncommon to see the instrument played with a pair of sticks in each hand, with notes alternately struck. The instrument is of Latin origin and is rarely used in symphonic writing.

The xylophone may or may not have resonators, depending on its size. The tone is very crisp if hard mallets are used, and there are often pronounced nonharmonic overtones in the sound. Notes are often rolled for a sustained effect.

The nonpitched xylophones include the following.

Woodblocks are made of hollowed-out sections of hard wood that may be struck with a drumstick, producing a hollow, snapping sound. They have been widely used in jazz and occasionally in orchestral music for special effect.

Castanets are small cup-shaped sections of hard wood that can be clapped together, producing a light, high-pitched sound. Repetition can be very fast. It is distinctly Iberian in origin and is generally heard as an adjunct to Spanish music.

The slapstick consists of two pieces of wood hinged together. When they are rapidly brought together, a loud "clap" is produced. The sound is often associated with comic routines and is used for special effects.

Latin dance music makes use of a variety of instruments that are best categorized as xylophones. Maracas are small dried gourds with seeds inside. When shaken, they produce a high-pitched rattling sound. Claves are a pair of round hardwood sticks that are struck together. We can add to this list a variety of notched gourds that are played with a scraper.

8.5   RANGES OF THE PITCHED METALLOPHONES AND XYLOPHONES

The following list gives the normal ranges for the various pitched percussion instruments. The ranges are assumed to be chromatic.

Xylophone	C_5 to C_8
Marimba	C_3 to C_7
Orchestra bells	G_5 to C_8
Chimes	C_4 to F_5
Celesta	C_4 to C_8

8.6   SPECIAL EFFECTS

The percussion section (see figure 8-2) of the orchestra is normally called upon to handle any special effect, whether percussion or not. Some of the more interesting are the following:

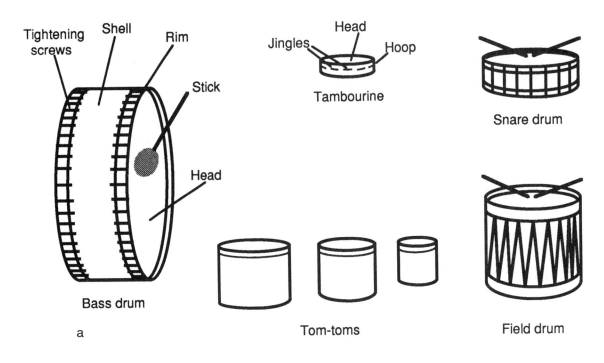

Figure 8-2. Various types of percussion instruments. (*a*) Membranophones; (*b*) orchestra bells; (*c*) the vibraphone; (*d*) the marimba. Photos courtesy of the Selmer Company.

c

d

The wind machine consists of a large circular structure with open metal meshwork on its cylindrical outer surface. A piece of canvas, or other suitable cloth, is draped over the mesh, and the circular part is cranked. The friction between the cloth and the metal mesh produces a swishing sound that rises or falls in pitch, depending on how fast the crank is turned. The sound is a little like that of a brisk winter wind.

Figure 8-3. The modern drum set.

Continuing in the weather department, the thunder sheet is a long, wide piece of sheet metal hanging from an overhead beam. When it is grasped at the bottom and shaken, the vibrations are somewhat reminiscent of thunder.

In 1924, Ottorino Respighi (1879–1936) wrote *The Pines of Rome*, in which a recording (called out by catalog number) of a nightingale was to be played at a specific point in the score. The job fell to the percussion section.

Today, many special effects are easily synthesized and played over loudspeakers with minimum difficulty.

8.7   THE DRUM SET

No discussion of percussion instruments would be complete without a description of the modern drum set. The drum set evolved out of early theater pit orchestras and quickly found its way into jazz groups. From there it spread to all fields of popular music and the rock music scene.

The drum set is intended to be played by one person, and that person is responsible for the rhythmic underpinning of the entire ensemble. Figure 8-3 shows a typical drum set. The actual componentry may vary from player to player, but certain elements are common to all. In the center is a bass drum, which is small by symphonic standards. The drum is played by a beater actuated by the right foot. To the player's left is a set of "high-hat" cymbals. This is a pair of cymbals, the bottom one fixed and the top movable by action of the player's left foot.

Closest to the player is a snare drum, which is played either with sticks or with wire brushes. Up to three tom-toms may be positioned in and around the components already described, and the space overhead in front of the player is occupied with various cymbals.

For the most part, the drummer improvises as he goes, following only the broadest guidelines if a musical score is presented to him. The drum set requires skills of coordination that at times seem awesome, and they are best learned at an early age.

The dynamics of the drum set can be extremely wide, as any studio recording engineer can attest. In the recording studio, microphones are often placed quite close to the drum set in order to achieve good isolation for precise assignment on the stereo sound stage. With close placement, many microphones may be required to ensure that all elements are present in the recording. The minimum number of microphones used is three: two in an overhead stereo configuration and one close to the kick drum. If more detail is desired in the pickup, individual microphones can be placed quite close to the high-hat cymbals, the snare drum, and each of the tom-toms. Small microphones are best to use, since they can be located without taking up too much space. Care must be taken that they are away from the range of motion of the drumsticks, and they must be placed in such a way that they pick up the desired sound. A microphone at the edge of a snare drum or tom-tom located 2 or 3 cm above the head will do nicely. Since cymbals move when they are played, the microphone must be located so that it does not fall into the null zone along the plane of the cymbal. A good position is under the cymbal facing upward. For internal microphone placement within the drum set, directional (cardioid) patterns are best. These details will be covered in chapter 13.

8.8   ACOUSTICAL CHARACTERISTICS OF PERCUSSION INSTRUMENTS

As a group, the percussion instruments have the widest dynamic range of the orchestral instruments. Any device can be struck very lightly or heavily, and the resulting output may vary by up to 80 dB or more. For example, a cymbal may be gently stroked with a light metal brush, or a pair of them may be struck together. The sound quality will be quite different in each case, and the level difference very high. For many percussion instruments, it can be said that there is no fundamental lower limit to how softly they can be played. Only local masking noise levels will effectively limit what the player can do.

At the other extreme, some percussion instruments can be played extremely loudly, depending basically on the strength of the player. Consider that a bass drum or a set of timpani can produce as much acoustical power output as the rest of a symphony orchestra! The reason, of course, is that the mechanical impedance match between the driving source (the player) and the load (the instrument) is fairly good. Then, if the load itself has large area, it couples effectively to the air, producing a large acoustical power output.

This is not to say that the overall process is truly efficient. A large physical gesture on the part of the player of a bass drum may expend many watts, while only a handful of acoustical watts will be radiated by the instrument. The fact is that the aerophones and chordophones are by comparison very inefficient.

Table 8-2. Output Capabilities of Some
Percussion Instruments

	SPL (1 m), dB		
	High:	Low:	Range:
Cymbal	116	50	66
Snare drum	117	60	57
Bass drum	122	43	79
Timpani	122	36	86

The data shown in Table 8-2 are adapted from Olson (1952). Here, the maximum output powers and sound pressure levels of several percussion instruments are given.

The directional properties of percussion instruments are generally quite complex. Such instruments as the tam-tam and the cymbal radiate at their lower modes essentially as dipoles, as discussed in section 1.9, producing maximum output perpendicular to the plates and minimum output in the plane of the plates. Thus, a suspended cymbal will radiate to the audience in a different manner than will a pair of cymbals that the player has just clashed together and that are both facing the audience.

The bass drum's lowest mode, that in which the enclosed air mass moves from side to side, is effectively radiated as a dipole; however, other modes are essentially omnidirectional. Recording engineers often ask the player to turn the drum so that one head faces outward toward the microphone array, in order to avoid partial cancellation of the lowest mode.

REFERENCES

Backus, J. 1969. *The Acoustical Foundations of Music*. New York: W. W. Norton.

Benade, A. 1976. *Fundamentals of Musical Acoustics*. New York: Oxford University Press.

Berg, R., and D. Stork. 1982. *The Physics of Sound*. Englewood Cliffs, NJ: Prentice-Hall.

Campbell, M., and C. Greated. 1987. *The Musician's Guide to Acoustics*. New York: Schirmer Books.

Morse, P., and U. Ingard. 1968. *Theoretical Acoustics*. Princeton, NJ: Princeton University Press.

Olson, H. 1952. *Musical Engineering*. New York: McGraw-Hill.

Pierce, J. 1983. *The Science of Musical Sound*. New York: W. H. Freeman.

Rossing, T. 1990. *The Science of Sound*. Reading, MA: Addison-Wesley.

Sachs, C. 1940. *The History of Musical Instruments*. New York: W. W. Norton.

ADDITIONAL RESOURCES

The New Grove's Dictionary of Music and Musicians. London: Macmillan, 1980.

The New Harvard Dictionary of Music. Cambridge, MA: Harvard University Press, 1986.

The New Oxford Companion to Music. New York: Oxford University Press, 1983.

9

Acoustics of Keyboard Instruments

For reasons having to do mainly with playing skills and techniques, keyboard instruments are generally discussed as a group. A celesta could just as easily be classed as a percussion instrument, as the keyboard actuates hammers that strike metal bars. The piano, harpsichord, and clavichord could be considered string instruments, since their strings vibrate and transfer power to a sounding board, as discussed in section 4.2. And of course the pipe organ could be considered a very large and complicated aerophone. Electronic keyboard instruments are in a class of their own and have been referred to by Sachs as *electrophones*. Since all traditional keyboard instruments have at least one sound source for each key, it is possible to play any number of notes at one time.

A keyboard player is often proficient on all these instruments, and the principal keyboardist in the symphony orchestra is expected to double, as required, on all of them.

Keyboard instruments are quite complicated mechanically, often requiring skills beyond those of the player for routine upkeep and tuning.

Our discussion begins with two historical instruments, the clavichord and the harpsichord. Then we move on to the piano, celesta, and pipe organ, concluding the discussion with electronic keyboard instruments.

9.1 HISTORICAL INSTRUMENTS

The clavichord is a relatively simple instrument. The back end of the key supports a small metal piece, called a *tangent*, that strikes a transversely mounted string. When struck, the string emits a sound, which is then transmitted through the bridge to the sounding board. When the key is released, the string is quickly damped by a piece of cloth, called *listing*. The basic mechanism is shown in figure 9-1.

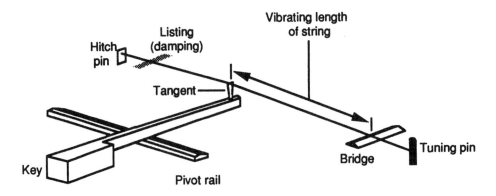

Figure 9-1. Simplified view of the action of the clavichord.

The clavichord is a small instrument, and its compass is normally no more than about four octaves. The instrument is relatively portable and may sit on a table of suitable height for playing. The sound output is quite small and does not project well; however, the instrument has some dynamic range capability. It is unique among keyboard instruments in that the player's finger remains in contact with the string (through the tangent) as the note is played and held. A variation in the player's touch can produce a small change in the string tension, and a slight degree of vibrato, called *bebung*, is possible. The clavichord was widely used as a practice instrument in early times.

The harpsichord is a more complex instrument in which the motion of the key is transmitted to a vertical member, called a *jack*, which plucks the string. As shown in figure 9-2, the string runs from front to back. There are often multiple sets of strings, each set controlled by its own jack slide, which is in turn actuated by a stop knob on the front of the instrument. The jack slide allows the player to engage one set of strings, as desired. When the key is released, the jack falls back, and a damper at the top of the jack mutes the string. Harpsichords have multiple sets of strings, or stops, at different pitches (unison and octave relationships) and of different timbres. The purpose is to allow the player variety within a texture that has fixed dynamics.

In overall shape the harpsichord resembles a small grand piano. The instruments often have two keyboards, with multiple stops (sets of strings) per keyboard. Some have been built with pedal sections, allowing them to play certain organ literature. The sound of the instrument is robust and projects well because of the presence of many high harmonics in the tone. The player cannot control the volume level by touch, inasmuch as the plucking action of the jack depends on upward displacement rather than velocity.

9.2   THE PIANO

Bartolomeo Cristofori (1655–1731) is generally credited with the invention of the pianoforte in the early eighteenth century. The name of the instrument means "soft-loud" in Italian and of course refers to the instrument's ability to play at many dynamic levels. In time, the name simply became piano, but the term *fortepiano* still lingers as a description of some early instruments.

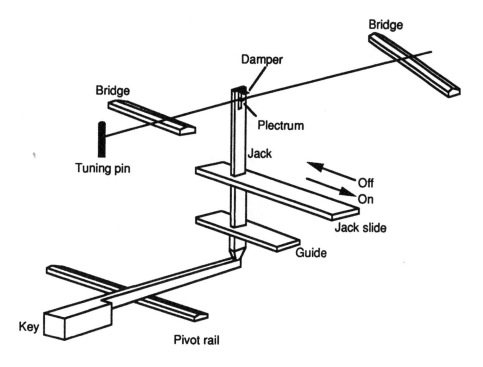

Figure 9-2. Simplified view of the action of the harpsichord.

The modern concert grand piano has evolved essentially as a solo virtuoso instrument. It is capable of considerable acoustical power output, mainly as a result of its size and high-tension stringing. The action is capable of rapid articulation for both soft and loud playing, yet it remains relatively light to the touch [about 50 g (1.8 oz) in the middle and upper ranges].

9.2.1   Details of Action

The modern grand piano action is shown in figure 9-3. When a key is struck, the felted hammer is thrown upward toward the strings. After striking, the hammer falls back to an intermediate position, some 10 to 12 mm (0.4 to 0.5 in) below the string. From this position, the hammer can again be thrown against the string if the player strikes the key quickly in succession. Since the hammer does not have to return to its rest position between strokes, rapid articulation is possible. On a well-regulated action, notes in the middle and upper part of the compass can be repeated cleanly as many as ten times per second. The loudness of the tone depends on the velocity of the hammer stroke, and a normal maximum range of loudness in the mid and lower ranges of the instrument would be about 30 to 35 dB on single keystrokes. The range from very soft single note attacks to loud chordal attacks might well be another 10 dB greater.

An essential part of the action is the damper assembly. The felt dampers are normally in contact with the strings, resting on top of them. When a key is struck, the damper is raised; it is held in this position as long as the key is depressed. When the key is released, the damper returns to the strings to mute the sound.

Figure 9-3. Concert grand piano action. Photo courtesy of Steinway & Sons.

9.2.2 Strings and Sounding Board

The compass of the grand piano is normally from A_0 to C_8, although the Bösendorfer Imperial grand has a full eight-octave compass, from C_0 to C_8. Over most of its range (from about D_2 to C_8), each note uses three steel strings tuned in unison. The lower notes of the instrument use wound strings, whose extra mass allows tuning to lower pitches without excessive length. Most of these are also in multiples of two or three, but the bottom strings are single.

The strings are mounted on a cast iron frame, necessary to withstand the total string tension, which can be as high as 18,000 kg (about 20 tons) force. The bass strings are mounted above the treble strings and splayed inward in what is called *cross-stringing* or *overstringing*.

The strings are struck at a point about one-eighth to one-tenth their length, depending on the register of the instrument. At the end of the instrument away from the player, the strings firmly engage the bridge, which rests on the sounding board and imparts vibration to it. There are two bridges, one for the steel treble strings and a bass bridge for the overstrung portion of the instrument. The acoustical reason for overstringing is that it allows the bass bridge to engage the sounding board at a mid position more favorable for acoustical coupling.

Both a concert grand and a spinet piano have the same range of notes; however, the string length may vary considerably, especially at lower frequencies. The reason goes back to the factors determined by Mersenne, which were discussed in section 4.2. The pitch of a string is a function of its tension, length, and mass per unit length. In a smaller piano with shorter strings, the tension and mass per unit length are set so that a shorter overall length is sufficient to produce the desired pitch.

The sounding board itself is generally made of spruce and is about a centimeter (0.4 in) thick, tapering at its edges. The underside of the sounding board has a number of wood reinforcing bars whose acoustical effect is not unlike that of the bass bar in the violin.

A fine violin generally improves with time, but a piano does not. While the action and stringing can be updated at any time, the sounding board changes for the worse with age. In a new instrument the sounding board has a slight upward arch, or crown, which is necessary to counteract the downward thrust of the strings. Over a period of many years, the sounding board tends to flatten, and the tone of the instrument deteriorates. One could conceivably replace the sounding board, but it is probably more cost effective to buy a new instrument.

9.2.3   The Pedals

The modern grand piano has three pedals. The left pedal is referred to as the *una corda* (single-string) pedal. It shifts the entire action slightly to the right so that the hammers strike only two of the three strings, producing a somewhat muted tone with a slightly different decay pattern.

The middle pedal is a more recent invention and is called the sostenuto pedal. When it is actuated, any dampers that are in the raised position (off the string) will remain in that position. In performance, notes can be "set" with the sostenuto pedal so that they will remain undamped, regardless of the position of the key.

The right pedal is known as the damper pedal. When it is depressed, the entire set of dampers is raised, letting all strings ring out freely.

9.2.4   Tuning

While normally tuned in equal temperament, the piano requires some degree of "stretched" tuning at the extremes of its compass. The reason for this is that the harmonics in a piano string are not quite integrally related. Specifically, the upper harmonics are slightly sharp relative to the pure harmonic series. Piano strings are quite stiff, acting somewhat like bars rather than simple strings. In figure 9-4a, we show the ideal displacement of a string for its first two harmonics. The string is effectively hinged at each end, and the second harmonic is exactly twice the first. In figure 9-4b, we show, in exaggeration, the effect of inharmonicity in a piano string. It is too stiff to be considered hinged at its end points, and the actual portion that vibrates is somewhat shorter than the entire length. As a result of this, the second and higher harmonics engage progressively less of the string, and the pitch of the higher harmonics is slightly higher than in the ideal case.

In tuning the instrument, the fundamental of the string an octave above is tuned so that it is beat-free with the second harmonic of the string an octave below. A good tuner does this strictly by ear, inasmuch as the amount of stretching will vary between instruments. Figure 9-4c and d shows the degree of stretching that may be encountered in a concert grand and a small upright piano. Note that the degree of inharmonicity between C_4 and C_6 may be only in the range of 10 or so cents. The reason for the increased inharmonicity in the upright piano is that the lower strings, because of their shorter length and increased mass per unit length, are significantly stiffer than the corresponding strings on the concert grand. As a result, they behave more like bars than the corresponding strings of the concert grand.

A matter somewhat related to tuning is the sustaining ability of the instrument. A piano note exhibits a double decay characteristic, decaying rapidly at first and then more slowly. At close quarters on a good instrument, a loudly struck tone in the range of C_4 can still be heard some 20 sec later, while a tone in the range of C_2 can be heard some 45 sec later. If the damper pedal is pressed, these times may be longer still.

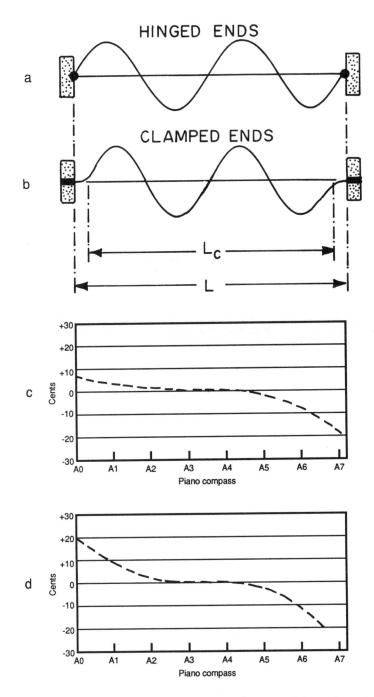

Figure 9-4. Stretched tuning of the piano. (a) Vibration of string with hinged terminations for first and second harmonics; (b) vibration of semirigid piano string for first and second harmonic. Charts showing the degree of stretched tuning (c) in a concert grand and (d) in an upright piano indicate the variation in cents that results from stretched tuning. a and b from Benade, 1976. Courtesy of Virginia Benade.

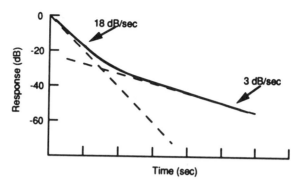

Figure 9-5. Characteristic double decay of a piano tone.

The reason for the double decay characteristic is that the bridge provides less mechanical impedance for string motion perpendicular to the sounding board than it does for string motion parallel to the sounding board. When a note is struck, the trio of strings executes essentially up-and-down motion, perpendicular to the sounding board. The strings are never exactly in unison and tend to couple and influence each other. They are never struck with a motion that is absolutely perpendicular, and the motion is slightly elliptical, containing a perpendicular component and a sidewise component parallel to the sounding board. With the higher mechanical impedance for the motion parallel to the sounding board, power is transferred from the strings at a lesser rate, and the decay rate is longer. Thus, the vertical motion of the strings dies out faster, and in the end, the motion is essentially from side to side. To some extent, a good tuner can influence this complex process by how much he may choose to "detune" the set of strings. A graph showing the typical decay characteristic of the piano is shown in figure 9-5.

9.2.5   The "Prepared" Piano

Over the years, *avant garde* composers have altered the instrument for their own purposes, often in a violent way. Musical scores have called for coins, screws, bolts, rubber erasers, and the like, to be wedged between strings for the sake of creating new sounds. Detunings are common as well. While some of these treatments may be benign enough, others are sufficient to break strings, or to bend them permanently. No good instrument should be subjected to such abuse.

9.2.6   Special Playing Techniques

The normal piano technique involves scale and chordal writing in which each finger strikes a single note. Tone clusters are groups of notes that may be played by the forearm, or with a stick, as may be indicated by the composer. The glissando is a rapid drawing of the thumbnail over a large range of either black or white keys.

In a technique reminiscent of triple tonguing, notes may be repeated rapidly by alternating the fingers on a single key. Half pedaling is an important technique in music of the Romantic and Impressionistic periods. It involves a partial action of the damper pedal to half mute the previous notes, allowing them to sound at a softer level. The sostenuto pedal may be used to set certain strings so that they will sound, not by striking, but solely through sympathetic vibration. In rare cases, the player may be asked to reach inside the instrument and sound the strings by hand.

9.3 THE CELESTA

Invented by Victor Mustel (1815–1890) in 1886, the celesta consists of a set of metal bars mounted over wooden resonators and struck by felted hammers. The action is not subtle, and the dynamic range of the instrument is rather limited, at least as compared with the piano. The range is normally from C_4 to C_8. The instrument is known to most people primarily for its important role in the Dance of the Sugar Plum Fairy from Tchaikovsky's (1840–1893) *Nutcracker* ballet, but it has been widely used in modern scores by Richard Strauss (1864–1949), Igor Stravinsky (1882–1971), and Béla Bartók (1881–1945).

The action is often noisy and difficult to maintain. Recording engineers have learned not to get too close if they are to avoid action noises. The celesta is shown in figure 9-6.

9.4 THE PIPE ORGAN

Unquestionably, the pipe organ is the largest and most complex of all keyboard instruments. While a few instruments are more or less portable, and useful in continuo playing (see section 10.1), most organs are firmly fixed in a given location, and are designed and voiced for the specific acoustics of that space.

The instrument goes back to antiquity, but we will discuss it only in terms of present requirements. Until about the middle of the nineteenth century, most organ actions were mechanical, with direct force by the player's finger opening the valves to let air into the pipes. Later, arrangements such as the Barker lever made use of air-assisted valve pulldowns, and eventually electromagnetic arrangements were used for this. Figure 9-7 shows these three types of action.

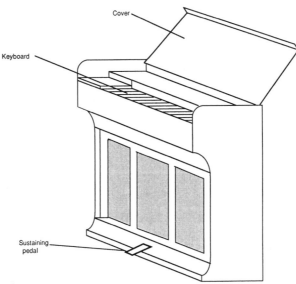

Figure 9-6. The celesta.

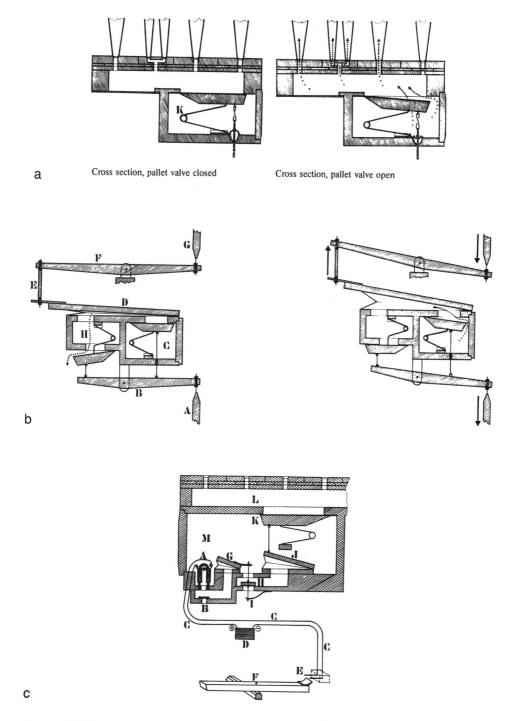

a Cross section, pallet valve closed Cross section, pallet valve open

b

c

Figure 9-7. Various pipe organ actions. (a) Tracker; (b) the Barker lever; (c) electric action. Data from Sonnaillon, *King of Instruments*, courtesy of the Rizzoli Company, New York.

9.4.1   Organ Pipes

An organ pipe sounds only one note, so the full manual compass of sixty-one notes requires as many pipes for each stop. Such a group of pipes is referred to as a *rank* of pipes and is controlled by a stop knob accessible to the player.

There are basically two types of organ pipes. Flue pipes resemble the flute family in that a jet of air is aimed at an opposing edge. Air flow causes the pipe to sound, in accordance with Bernoulli's principle. Flue pipes may be made of metal or wood, and they may be stopped or open.

Reed pipes consist of a reed assembly and an associated resonator. The reed itself can be tuned to a given note, with or without the resonator. A short resonator can then be used to emphasize certain formants in the reed tone or, if of the right length, to reinforce the entire harmonic series of the reed. In the latter case, the pipe may be thought of as a double system in which both reed and resonator are tuned to the same frequency. Such pipes can remain quite stable in pitch, assuming that the temperature does not vary substantially.

Most organ stops consist of one pipe per note; however, mixture stops, which sound various octave and fifth combinations at higher pitches, may consist of three to six pipes per note.

Figure 9-8 shows the profiles of various organ pipes.

9.4.2   The *Werkprinzip*

Most instruments built today, whether for liturgical or concert hall use, are designed along the German *Werkprinzip*, or work principle. This notion of organ building was brought to full bloom during the Baroque period, and it defines each manual or division of the organ as an individual instrument in its own right. Thus, each division will have a complete ensemble of individual ranks of pipes and mixtures playing at a variety of pitches related to the harmonic series.

Stop pitch nomenclature is based on the length of an open pipe that will sound C_2 when that note is pressed on the manual keyboard. The pipe length necessary to do this is 8 ft. Thus, an 8-ft rank of pipes sounds the pitch as written and is often referred to as of unison pitch.

A 4-ft pipe sounds an octave higher, and so forth. The following list indicates the pitches of the various speaking lengths of organ stops.

Pitch length	Sounding
32 ft	Two octaves lower
16 ft	Octave lower
8 ft	Unison
4 ft	Octave above
2 2/3 ft	Twelfth above
2 ft	Two octaves above
1 3/5 ft	Seventeenth above
1 1/3 ft	Nineteenth above
1 ft	Three octaves above

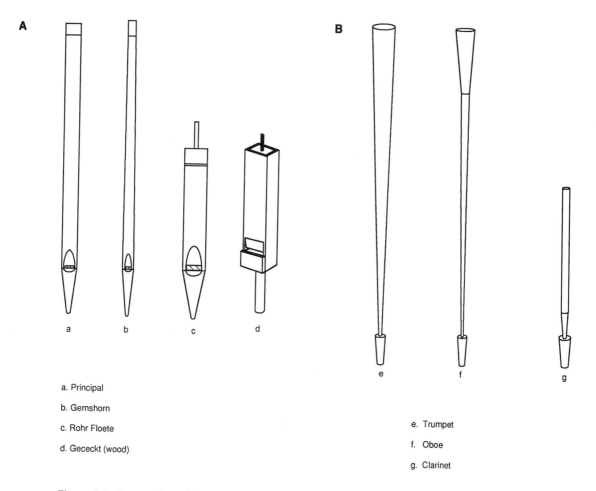

A

a
b
c
d

a. Principal

b. Gemshorn

c. Rohr Floete

d. Gececkt (wood)

B

e
f
g

e. Trumpet

f. Oboe

g. Clarinet

Figure 9-8. Organ pipes. (*a*) Representative flue pipes; (*b*) representative reed pipes.

The written manual compass is normally from C_2 to C_7, and there are usually two to three manual divisions. The written compass of the pedal division is C_2 to G_4. If the instrument contains a 32-ft stop in the pedal division, it will sound the note C_0 when the lowest pipe is sounded.

Figure 9-9 shows the layout of a four-manual instrument embodying the *Werkprinzip*. Note that each division of the instrument is given its own position in the organ case. In most modern instruments, at least one division is enclosed in a wooden "box" with a set of louvers, called *swell shades*, that can alter the loudness of that division. The swell shades are operated by a pedal at the playing desk.

Within the *Werkprinzip*, each manual or division of the instrument fulfills a specific musical function. The *Hauptwerk*, or Great division, is the most assertive and is used for primary musical exposition. It contains both narrow- and wide-scale flue stops at various pitches and can build from a moderate level up to a full level, or *plenum*, depending on stop registration. Reed stops of the trumpet family provide a capstone for the division. The secondary division is called the

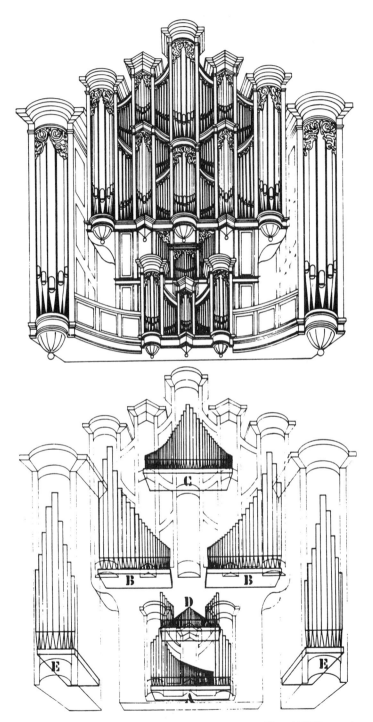

Figure 9-9. Division layout of a four-manual organ. From Sonnaillon (1985), courtesy of the Rizzoli Company, New York.

Positiv division and is somewhat lighter in texture, providing a plenum of its own, along with a variety of coloristic stops. The modern instrument has its Swell division, enclosed, providing a rich array of reed stops and softer stops of accompanimental nature. The Pedal division has the variety to support all these functions. Over the years, performance tradition, along with direct indication by composers, has indicated what basic registration and manual functions are to be used.

9.5   ELECTRONIC KEYBOARD INSTRUMENTS

There are various kinds of electric pianos, the simplest using small bars with a mechanical action and damper assembly. The tone from the bars is amplified and perhaps further subjected to tone enhancement before being fed to an amplifier-loudspeaker chain. The sound only superficially resembles that of a piano, and such instruments are used mainly in popular and rock ensembles. Later models are all electronic, with touch-sensitive keyboards.

Electronic organs date back to the 1930s. While early models bore only a faint resemblance to the real thing, modern electronic organs very often fool the ear. For many liturgical applications, they represent the only economic choice. Even then, they can be as expensive as the buyer may wish. There are a few examples of combinations of pipe ranks and electronic tone generation, representing some sort of middle ground between two often opposing philosophies.

9.6   ACOUSTICAL CHARACTERISTICS OF KEYBOARD INSTRUMENTS

9.6.1   Power Output

While the harpsichord has quite low power output, its tone carries well above moderate size ensembles because of its rich harmonic development. The instrument has a cover, which is usually open at an angle of about 45°, thus projecting the sound outward. For recording purposes, the cover may be removed.

The concert grand piano can produce peak levels at a distance of about 1 m (3.25 ft) in the range of 105 to 110 dB, corresponding to a power output in the range of 0.5 W. In concert performance, the instrument's cover is full up, at an angle of about 45°, projecting the sound outward toward the audience. When used for accompaniment, the instrument is normally played on "half stick," that is, raised only about 20 cm (8 in). Under this condition, the instrument sounds softer and muted. In the orchestra, the instrument is often played with its cover removed.

The power output of the pipe organ depends on the size and character of the instrument. Organs designed along historical principles are not as loud as the instruments that were built half a century ago, and it is no exaggeration to state that some of the louder instruments of the first half of this century could produce power outputs in the tens of acoustic watts. Certainly their power output could exceed that of a symphony orchestra. In terms of overall dynamic range, the organ can exceed all other instruments or combinations of instruments.

9.6.2 Directional Characteristics

Figure 9-10 shows the variation in output from a concert grand for four microphone positions close to the instrument. Note that the variation at lower frequencies can be as much as 30 dB. For lower frequencies, the sounding board acts as a dipole, and recording positions such as 1 and 3, which are more in line with the sounding board, will pick up less bass than will positions 2 and 4.

Figure 9-11 gives another view of the same condition. Here, two different recordings were analyzed with one-third octave peak hold. The music was similar, as were playing levels. In one case, the cover of the instrument was removed and the microphones were placed overhead. In the other case, the cover was in place and the microphones were placed more in line with the sounding board. Low-frequency cancellation is apparent in the recording made with the microphones in the dipole null zone. This latter case is in fact the way the instrument is normally heard in a concert hall.

Sound from the pipe organ normally reaches the listener via indirect paths, because of the interference of the organ case. The pipes themselves are virtually omnidirectional in their radiation, inasmuch as the mouth and open end of the pipe are small with regard to the fundamental frequency. For very high harmonics, preferential radiation is normal to the mouth and along the axis of the opening at the top.

In some organs, a rank of reed pipes may be placed horizontally, with the bells facing out into the room. As with trumpets in the orchestra, the highest harmonics of these pipes will be quite directional along their axes.

Traditionally, organ builders have voiced instruments so that the balance between lower, middle, and higher frequencies compensates for the indirect radiation of the instrument and the effects of room reverberation.

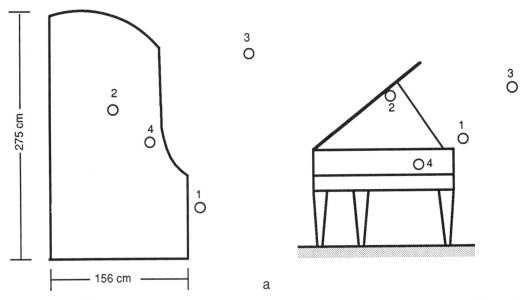

Figure 9-10. Variation of output from a concert grand piano. (*a*) Microphone locations; (*b*) tonal variation.

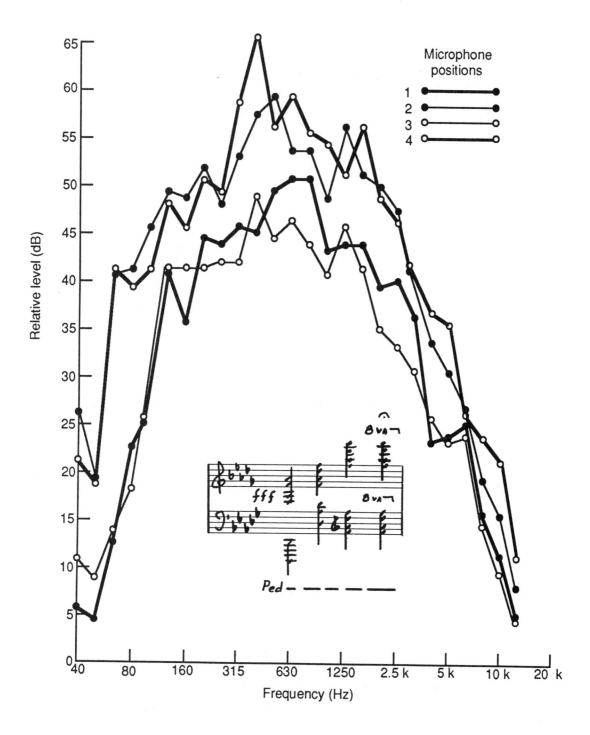

b

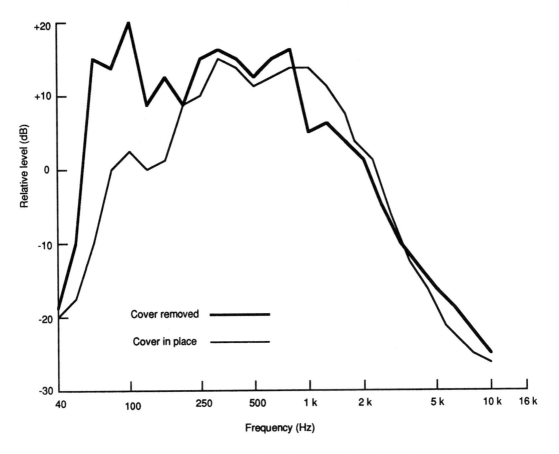

Figure 9-11. Recorded peak hold spectra of two piano recordings. Heavy line—cover removed; lighter line—cover in place.

REFERENCES

Andersen, P. 1969. *Organ Building and Design.* New York: Oxford University Press.

Audsley, G. 1905. *The Art of Organ Building.* Reprint. New York: Dover Publications, 1965.

Backus, J. 1969. *The Acoustical Foundations of Music.* New York: W. W. Norton.

Benade, A. 1976. *Fundamentals of Musical Acoustics.* New York: Oxford University Press.

Berg, R., and D. Stork. 1982. *The Physics of Sound.* Englewood Cliffs, NJ: Prentice-Hall.

Campbell, M., and C. Greated. 1987. *The Musician's Guide to Acoustics.* New York: Schirmer Books.

Culver, C. 1956. *Musical Acoustics.* New York: McGraw-Hill.

Gill, D. 1981. *The Book of the Piano.* Ithaca, NY: Cornell University Press.

McMorrow, E. 1989. *The Educated Piano*. Edmunds, WA: Light Hammer Press.

Moravcsik, M. 1987. *Musical Sound*. New York: Paragon House Publishers.

Olson, H. 1952. *Musical Engineering*. New York: McGraw-Hill.

Pierce, J. 1983. *The Science of Musical Sound*. New York: W. H. Freeman.

Rossing, T. 1990. *The Science of Sound*. Reading, MA: Addison-Wesley.

Sachs, C. 1940. *The History of Musical Instruments*. New York: W. W. Norton.

Sonnaillon, B. 1985. *King of Instruments*. New York: Rizzoli.

ADDITIONAL RESOURCES

The New Grove's Dictionary of Music and Musicians. London: Macmillan, 1980.

The New Harvard Dictionary of Music. Cambridge, MA: Harvard University Press, 1986.

The New Oxford Companion to Music. New York: Oxford University Press, 1983.

10

Musical Ensembles

In this chapter we will deal with musical ensembles, briefly tracing their history from the beginnings to the present day. We will discuss acoustical balances within the ensembles, traditions in seating, and acoustical output levels.

We will evaluate the changing performance requirements brought on by today's larger venues and discuss the problems in intonation and rhythmic control that these larger venues often create. We will also discuss sound pressure levels within ensembles, pointing up some of the potential hearing problems that can result from overexposure to high levels.

10.1 EVOLUTION OF CHAMBER ENSEMBLES

The term *chamber music* describes almost any musical ensemble from solo instrument with piano up to a thirty-five- or forty-piece ensemble. Most of the traditional chamber ensembles had their beginnings in the middle to late eighteenth century and became more formalized as the nineteenth century got underway.

The *continuo* was an essential part of most ensembles of the Baroque era (roughly 1600 to 1750). The term refers to a musical function normally fulfilled by a single keyboard instrument, with a cello or bassoon doubling the bass line. While the bass line was clearly indicated by the composer, the keyboard part was largely improvised by the player, following harmonic indications (called figured bass) notated by the composer. The tradition is still carried on today in performances of Baroque music. The trio sonata was a popular chamber music form during the Baroque period; it consisted of two solo instruments supported by a continuo.

In time, the continuo declined, and the chamber music forms that developed during the so-called Classical period (1750 to about 1830) did not include it. The most important chamber ensemble to arise during the Classical era was unquestionably the string quartet, an outgrowth of the trio sonata. In common with the old form were the two treble parts, here played by violins, and the cello bass line. The addition of the viola complemented the spectrum in the middle register and increased the harmonic possibilities.

The typical seating for the string quartet as seen from the audience is, from left to right, first violin, second violin, cello, and viola, with the second violin and cello slightly behind the others. Seating can, of course, vary from this. The players are normally located in a rather tight arc so that they can maintain good contact through peripheral vision. In musical matters, the ensemble relies on collegial consensus, but behind the scenes things may not always be so simple. Acoustically, the string quartet has the advantage of near perfect balance within its parts. As we noted in chapter 5, the viola and the cello are both scaled smaller than the violin, in terms of their frequency range, and this provides a relatively lean sound in the middle and low voices that allows all parts to be heard with clarity. One can certainly wish for a bigger viola or cello tone for solo purposes, but in the quartet ensemble those instruments are justly balanced. The dynamic range of the quartet is little more than that of any of its constituent parts, and normal program ranges rarely exceed 30 to 35 dB. Unquestionably, Beethoven (1770–1827) brought the string quartet to its first musical heights, and the ensemble has remained popular well into modern times.

The increasing dominance of the piano over the harpsichord during the Classical period brought a number of chamber ensembles into being. The most important of these was the piano trio, consisting of violin, cello, and piano. As a rule, the piano is played on "half stick"—that is, with the cover only partly up—to provide more appropriate acoustical balance with the two string instruments. As seen from the audience, the piano is normally in the middle, with the violin at the left and the cello at the right, both in front of the piano.

The piano quartet consists of piano with three string instruments, normally violin, viola, and cello, while the piano quintet consists of piano plus string quartet. In either case, the instrumentation can vary slightly. In these ensembles, the piano is always located toward the rear, with the string instruments occupying their normal positions, high on the left to low on the right. The string quintet normally consists of a string quartet with an additional viola. Instrumentation may vary.

From the very beginnings of musical composition, composers have written for available resources, and this has enriched the chamber literature through the years. Many woodwind instruments were included with strings in chamber ensembles during the Classical era and well into the nineteenth century. In terms of wind and brass resources alone, the wind quintet has remained popular, consisting of flute, oboe, clarinet, bassoon, and horn. There have been many brass ensembles, most of them ad hoc combinations for the occasion or resources at hand. Most common is the brass quintet, consisting of two trumpets, horn, tenor trombone, and either bass trombone or tuba.

Today, most chamber music is composed not for traditional groups but for whatever ensemble appeals to the composer purely in structural and coloristic terms. The chief exception is the string quartet, whose appeal has been virtually constant since its inception.

Finally, let us observe that the simple sonata for solo instrument and piano has been one of the most enduring chamber forms, embracing over the years just about all possible instruments.

10.2   EVOLUTION OF THE ORCHESTRA

10.2.1   Terminology

When we speak of the strings of the orchestra, we are referring to an ensemble of violins (divided into firsts and seconds), violas, cellos, and basses. The number in each section may vary, but it is expected that there will be several players in each section, making a total of no less than fifteen or twenty, and possibly as many as sixty. The division of violins into firsts and seconds is purely for reasons of composition; the instruments are the same.

The ensemble of woodwind players is usually referred to simply as "winds," while the trumpets, trombones, and tuba are referred to as "brass." The French horns, though technically brass instruments, are usually categorized separately as "horns." The primary reason for this is that their musical function within the orchestra is often different from that of the other brass instruments. Specifically, their sound is radiated to the right rear of the player, and therefore reaches the audience through a reflected path. By comparison, trumpets and trombones radiate their sound directly toward the audience. In many musical works, the horns are used as a foil to the more forward brass instruments, and they are frequently used to augment the winds.

The percussion instruments are referred to as a group, with the exception of the timpani, which maintain a separate musical function in the orchestra.

10.2.2   Historical Evolution

The origins of the orchestra go back to the larger Baroque ensembles. The orchestra as we think of it today, however, did not take form until the Classical era, at which time it consisted of a fairly well defined string ensemble with variable winds, brass, and percussion. A study of the instrumentation of the symphonies of Franz Joseph Haydn (1732–1809) and Wolfgang Amadeus Mozart (1756–1791) will show how different the requirements could be, depending on musical needs and player availability. Generally, we can describe the minimal requirements as strings, winds and horns in pairs, with a pair of trumpets and timpani optional.

Normal seating called for divided violins, with the firsts on the left and the seconds on the right. Cellos and basses were normally behind the first violins, and the violas were behind the second violins. The winds were seated in the middle behind the strings, with horns and brass either to the right or to the left of the winds. Timpani were at the back. Details varied, but from the beginning the strings were placed at the front because of their somewhat weaker acoustical output. Figure 10-1 shows seating details of the chamber orchestra as it might be deployed today.

The rise of the so-called Romantic era in the early 1800s brought with it the large orchestra. Hector Berlioz (1803–1869), through his own music and his famous treatise on orchestration, gave good focus to the large ensemble that we know today as the symphony orchestra. The string complement numbered perhaps fifty-five or sixty, and winds were used in fours (sixteen players). Doublings were common, with the third and fourth players taking on the added instruments. Doublings in the flute section were on the piccolo, and those in the oboe section, on the English horn. Clarinet doublings were on the bass clarinet and the treble form of the instrument, and bassoon doublings were on the contrabassoon. The brass and horn players normally did not double, because of their relatively small number.

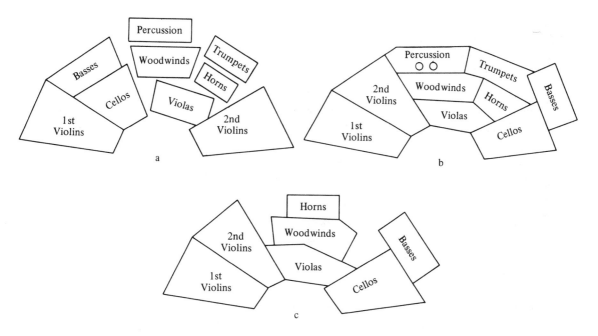

Figure 10-1. Typical seating arrangements for the chamber orchestra. (*a*) Older seating arrangement; (*b*) modern seating arrangement; (*c*) alternative seating for reduced resources. From Eargle 1986.

By the time Wagner (1813–1883) had adapted the orchestra to his specific operatic needs, the brass complement had grown considerably, and a typical requirement might be as follows: four trumpets, four trombones, one tuba, and eight horns (with players five through eight doubling on so-called Wagner tubas).

The large orchestra of today was only slightly modified by the demands of Gustav Mahler (1860–1911), Richard Strauss (1864–1949), and Stravinsky (1882–1971). Essentially, more flexible instruments were required, as were greater playing skills, and the conventional ranges of even the traditional instruments were forcibly extended by composers, often beyond traditional limits. The percussion section of the orchestra was amply augmented with larger, and certainly louder, instruments.

Table 10-1 lists the respective requirements of the Classical orchestra of Beethoven (1770–1827), the Romantic orchestra of Brahms (1833–1897), and the modern orchestra of Strauss (1864–1949).

10.2.3   Orchestral Seating

Prior to the early part of this century, the arrangement with divided violins shown in figure 10-2 was typical. It is important as well to note the seating within each section. Each instrumental section has a principal player, the player who normally plays solo passages when called for. In the string sections, all principals sit at the front of their respective sections. Therefore, the frontmost stands of players include the principals in a quasi-string quartet configuration. The bass principal is normally at the frontmost position of the row of basses.

Table 10-1. Typical Orchestral Requirements of the Classical, Romantic, and Modern Periods

Classical	Romantic	Modern
6 1st Violins	10 1st Violins	14 1st Violins
6 2nd Violins	10 2nd Violins	14 2nd Violins
4 Violas	8 Violas	10 Violas
4 Cellos	8 Cellos	10 Cellos
2 Basses	6 Basses	9 Basses
2 Flutes	2 Flutes	3 Flutes (doubling alto flute)
2 Oboes	1 Piccolo	1 Piccolo
2 Clarinets	2 Oboes	3 Oboes
2 Bassoons	1 Cor Anglais	1 Cor Anglais
2 Horns	2 Bassoons	3 Bassoons
2 Trumpets	1 Contrabassoon	1 Contrabassoon
1 Timpani	4 Horns	4 Horns
	3 Trumpets	3 Trumpets
	3 Trombones	3 Trombones
	1 Tuba	1 Bass Trombone
	1 Timpani	1 Tuba
	2 Percussion	1 Timpani
		3 Percussion
		2 Harps
		1 Keyboard

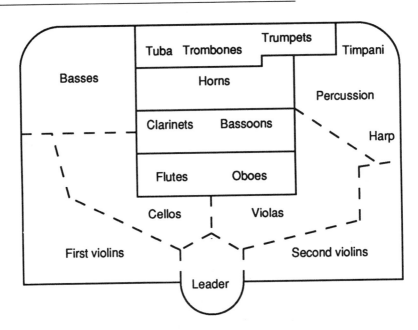

Figure 10-2. Symphony orchestra seating with divided violins.

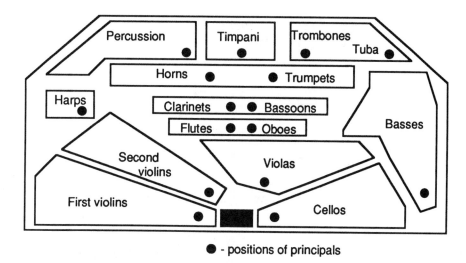

Figure 10-3. Symphony orchestra seating with all violins on the left. Dots indicate positions of principal players.

In the wind section, as seen from the conductor's position, there are two rows of players. In the first row, the flutes are at the left and the oboes are at the right. In the second row, the clarinets are at the left and the bassoons are at the right. The principals sit in the middle of each row, so that the four wind principals are clustered together, two in front and two in back. The advantage here is that they can easily confer during rehearsal in matters of sectional balance and intonation. And, of course, their ensemble playing during soli passages will be helped by their proximity. Similar considerations hold in the brass and horn sections of the orchestra.

There is a principal percussionist, and his main job is to establish balances and assign doublings, which can be quite complicated in modern scores. The timpanist is also a principal player and not, strictly speaking, part of the percussion section. Other principal players in the orchestra are the keyboardist and the first harpist.

Beginning in the early part of this century, the English conductor Henry Wood (1869–1944) reseated the strings in a high-to-low arrangement from left to right, as shown in figure 10-3 (Ranada 1986). This so-called new arrangement was used sporadically by a number of conductors, including Leopold Stokowski (1882–1977). But it was not until after World War II that the new seating became dominant in many parts of the world, chiefly in the United States. Wood had earlier stated that the commonality of writing for the first and second violins dictated that they be closer together for better ensemble. The same, of course, could be said for the violas and cellos. There was undoubtedly a good bit of truth here, but the factor that literally forced the new seating arrangement on most large orchestras was the desires of record companies during the 1950s. To them, the new arrangement simply sounded better in stereo, because of the clearer delineation of highs to the left and lows to the right.

Many arguments can be made for and against the two seating arrangements, and each conductor must make up his own mind. For most works of the Classical era, there is no question that the counterpoint between first and second violins is easier to hear when the

which is usually heard as:

a

Figure 10-4. Musical excerpts showing spatial effect of divided violins. (a) Tchaikovsky, Symphony No. 6, final movement, measures 1–2; (b) Stravinsky, *Scherzo Fantastique*, measures 94–96.

sections are divided. Such counterpoint, however, is not lost in the new seating arrangement any more than it is lost over a single-channel radio transmission. So the question remains moot.

There are, however, many passages in orchestral writing that in purely spatial terms may sound completely different, depending on violin seating. The most famous is perhaps the string passage that opens the fourth movement of Tchaikovsky's (1840–1893) Sixth Symphony (1893), as shown in figure 10-4a. In the new seating, or over a single-channel broadcast, the ear will simply hear the top note of each violin part as a continuous diatonic melody. Given the benefit of spatial separation, either in the concert hall or on a stereo recording, the effect is that of the melody being slowly tossed from one side to the other. Not many people have had the advantage of hearing it in this manner.

A second example, shown in figure 10-4b, is taken from Stravinsky's (1882–1971) *Scherzo Fantastique* (1908). Here, the "chase" between the two violin sections is apparent only if they are divided.

It is difficult in these examples to say whether the composers were writing purely for spatial effects or for other reasons having to do with ease of performance or matters of phrasing. In any event, the effect with divided violins is a striking one. One can also, with equal validity, point to examples using the new seating where spatial effects between high and low strings are noteworthy.

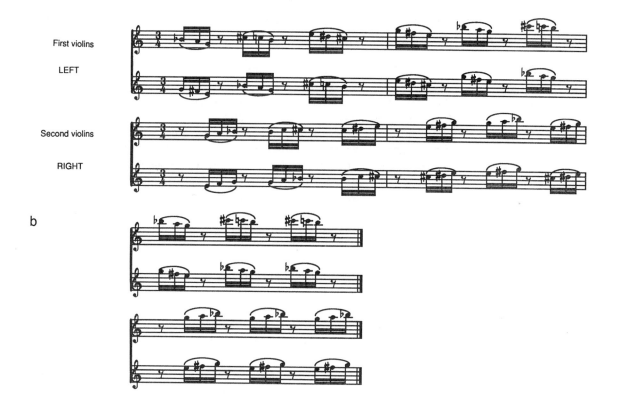

First violins

LEFT

Second violins

RIGHT

b

10.3   ACOUSTICAL POWER OUTPUT OF THE ORCHESTRA

10.3.1   Musical Dynamic Indications

The following terminology from Italian is generally used in music notation to indicate playing levels:

pp	*pianissimo* (very soft)
p	*piano* (soft)
mp	*mezzopiano* (moderately soft)
mf	*mezzoforte* (moderately loud)
f	*forte* (loud)
ff	*fortissimo* (very loud)

We occasionally see the terms *ppp* and *fff*, indicating further extremes in dynamics.

These terms are all relative, and the differences between them are not at all uniform from one instrument or instrumental section to another. As we have seen in earlier chapters, some instruments are inherently capable of wider dynamic ranges than others, and it is clear that the meaning of these terms must be adjusted not only to instruments and instrumental sections, but also to the pitch range involved.

10.3.2 Power Measurements

The normal acoustical power output of the modern symphony orchestra is in the range of 1 W. At peak output, the power may reach the 10- to 20-W range.

Veneklasen (1986) performed a set of orchestral power measurements in the Seattle Opera House and arrived at the data shown in figure 10-5. When all these power levels are summed, the total is just about 1 W. Since the sections of the orchestra were each playing at what they considered a *mezzoforte* level, we can clearly observe the differences in what *mf* implies for each section. Presumably, if asked to play *ff* or even louder, the sections could add another 10 dB to their output levels, bringing the total acoustical power output up to 10 W or so.

Veneklasen further experimented with various seating arrangements within the orchestra shell at the Seattle Opera House in an effort to alter the orchestral balance to give more advantage to the strings. He had determined by measurements of scale models of the shell that sectional balances could vary as much as 14 dB, depending solely on location within the shell. Specifically, Veneklasen noted that elevated sources on stage were heard at higher levels by the audience. The reason for this is that, from elevated positions on stage, sound sources incurred less attenuation because of their larger angle of incidence with respect to the seating area. In addition, the careful adjustment of stage canopies created useful reflections, which were themselves less affected by low grazing incidence angles with respect to the audience.

The net result of the reseating experiments is shown in the graphs of figure 10-6. Most significantly, the output of the brass section has been reduced 6 to 7 dB in the midband, providing better balance with the winds and strings.

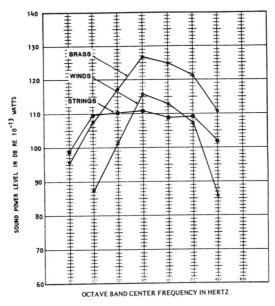

Figure 10-5. Power output levels of sections of the modern symphony orchestra. From Veneklasen 1986.

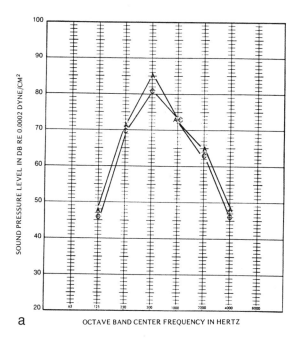

a

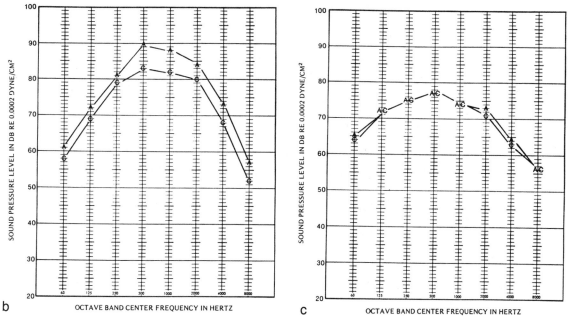

b   c

Figure 10-6. Effects on sectional sound pressure levels due to reseating of the orchestra. (*a*) Effect of reseating the woodwinds; curve A shows the response with the standard seating, and curve C shows the effect of reseating as in figure 10-7; (*b*) effect of reseating the brass; curve A shows the response with the standard seating, and curve C shows the effect of reseating as in figure 10-7; (*c*) effect of reseating the strings; curve A shows the response with the standard seating, and curve C shows the effect of reseating as in figure 10-7. From Veneklasen 1986.

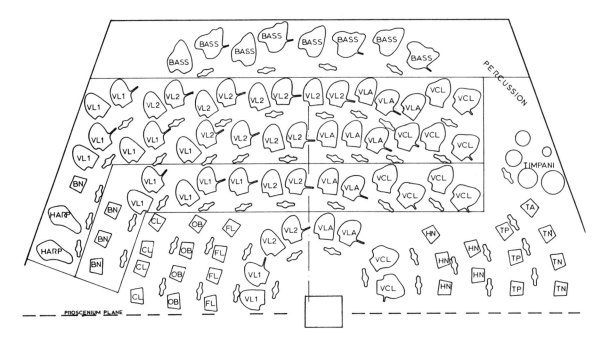

Figure 10-7. Veneklasen's experimental reseating of the symphony orchestra. From Veneklasen 1986.

The curves marked A represent the standard seating arrangement, with violins on the left, while the curves marked C represent an alternative seating arrangement shown in figure 10-7. Note that the strings have been placed toward the middle and rear of the stage, thus providing more reflected images off the overhead canopies. The brass and winds have been moved to the front of the stage, where there are fewer reflected images and where their locations result in lower grazing angles of incidence with respect to the audience.

It is truly impressive what can be done in these regards if the conductor is of an experimental turn of mind. Often, this is not the case, and balances continue as before. It is essential, also, that there be a traditional orchestra shell that is adjustable. In many concert halls there is no shell per se, and the orchestra performs essentially in an immediate environment that cannot easily be changed.

10.4   THE VOCAL CHORUS WITH ORCHESTRA

People have always sung and chanted, and the rise of formal liturgical chant during the Middle Ages established directions for the development of later forms. Simple unison singing gave way to multivoice composition, called *polyphony* (Greek for "many sounds"), in which each vocal section sang an independent melodic line in concert with the others. This style reached its zenith during the sixteenth century in the works of Palestrina (1525–1594) and Lassus (1532–1594).

Ultimately, vocal resources were integrated with instrumental ones, and the great traditions of church cantatas and oratorios developed, with emphasis on both choral and solo singing.

Most development of choral musical forms has taken place in liturgical settings in which long reverberation times have augmented, and often enhanced, the efforts of untrained singers. In contemporary concert performance, skilled singers are used in whatever number required for the work at hand.

Basically, there are six vocal ranges, as shown in figure 10-8. Not every voice fits neatly into these categories, and we often find singers whose capabilities encompass two or more of these ranges.

In a concert setting, the chorus is usually placed behind the orchestra. If there is an orchestra shell that tapers down toward the rear, then the chorus will be projected well forward. However, in many concert halls, the chorus performs without the benefit of close-in reflective surfaces. Figure 10-9 shows a plan view of Royal Festival Hall in London. Here, the chorus has an advantage in elevation, but there are no close surfaces to reflect the sound outward toward the audience. In cases such as this, very large choruses, often numbering well over 100, are needed for proper balance.

Choruses are normally divided into four sections, bass, tenor, alto, and soprano. As a rule, the men are at the rear and the women in front. There are many portions of works, old and new, with *divisi* passages, in which the sections must divide. Works with double chorus are common, and in these cases the groups often divide left and right for antiphonal effect. Some well-trained small vocal ensembles use mixed placement, in which the singers are placed more or less randomly. This practice yields a thoroughly homogeneous sound, often quite rich and "large." It is difficult, however, for the conductor to give cues, and the singers must be very competent and well rehearsed. It is worth noting that choruses always perform standing.

While many choral singers are trained as vocal soloists, they modify their singing technique for choral purposes. Blending of voices is of paramount importance, and singers in a chorus normally perform with less vibrato and overall volume level than do soloists.

Figure 10-8. Normal vocal singing ranges.

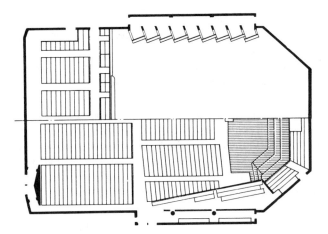

Figure 10-9. Section view of Royal Festival Hall, London. From Beranek 1962. Data courtesy of J. Wiley and Sons, New York.

10.5   ENSEMBLES IN THE OPERA HOUSE

As we shall see in section 11.2.1, the opera house, by virtue of its size, shape, and acoustical treatment, does not promote long reverberation times. It is essentially a theater in which singing, rather than speech, is the primary medium. Visual and dramatic elements dominate the stage, while the orchestra is placed out of sight in a pit in front of the stage, where it plays to little acoustical advantage.

For singers on stage, there is some benefit from early lateral reflections from the balcony fascia at various heights. The orchestra gains little from this, and in its pit location it assumes a role quite different from that which it has in the concert hall. The ensemble in the pit is quite large, and if it were on stage it could easily fill a concert hall; in the opera house it must somehow be throttled back to "make room" for the singers. Richard Wagner (1813–1883) designed the Festspielhaus in Bayreuth, Germany, for his mammoth-scale operas, and the large pit orchestra was acoustically scaled back by a combination of pit depth and baffling, as shown in figure 10-10. The large brass section was located well under the stage, and it was further shaded by the baffles. Only with the baffles in place could the brass instruments produce the tone that Wagner had in mind and not drown out the singers. Only the strings had easy egress into the hall.

The arrangement described above is more or less unique to Wagner's writing. The Bayreuth pit details have rarely been duplicated, and no other composer's writing fares well under such conditions. Conversely, it is often difficult to balance Wagner's orchestral and vocal resources in conventional opera houses.

In general, singers on stage are able to rise above a large pit orchestra through the following:

1. Orchestral/vocal balance and timing. Even though the orchestra may be louder than a singer, it rarely doubles the vocal line, and this gives some prominence to the singer. Furthermore, slight timing errors, such as the fact that the singer may anticipate a musical attack, will give further prominence (Sundberg 1982, 59–98).

2. Matters of vocal register. Low vocal registers are generally weaker than higher ones, and operatic composers have always taken this into account in the weight of orchestration.

3. The "singer's formant." Male singers cultivate a prominence in the 2- to 3-kHz range at high singing levels, which allows them to be heard easily against a full orchestral texture, as shown in figure 10-11.

4. Female singers can often modify formant frequencies to coincide with harmonics produced by their voices, thus producing a higher output.

Finally, a voice does not have to dominate and be "unmasked" at all times in order to be clearly heard and understood. It will suffice if it is in the clear only part of the time.

Operatic choruses on stage rarely number more than about sixteen. These singers must have strong voices of near-soloist projection if they are to be heard, inasmuch as they are usually positioned upstage and sing into an environment virtually devoid of beneficial reflections.

10.6   ENSEMBLES IN POPULAR MUSIC

Jazz was the earliest major influence on popular music, and it has traditionally relied heavily on improvisation. Thus, there are few formalized ensembles, and players tend to complement one another by skills rather than by instrument. One exception here is the so-called jazz big band, which makes use of elaborate arrangements as well as improvisation. The normal composition of the jazz big band is shown below.

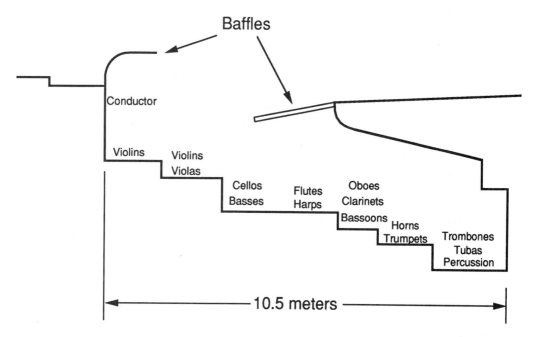

Figure 10-10. Section view of orchestra pit in Bayreuth Festspielhaus. From Beranek 1962. Data courtesy of J. Wiley and Sons, New York.

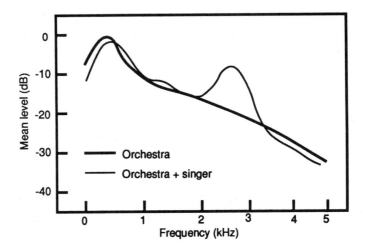

Figure 10-11. The "singer's formant." Spectra of orchestra and tenor with orchestra. From Sundberg 1982.

5 Trumpets	1 String bass
5 Trombones	1 Piano
2 Alto saxophones	1 Guitar
2 Tenor saxophones	1 Drummer
1 Baritone saxophone	

Of this basic ensemble, the bass, guitar, and piano have traditionally been amplified because of their relatively low acoustical output. The brass, saxophones, and percussion are capable of playing quite loudly and can balance themselves with little difficulty.

In modern arrangements there are often many doublings. For example, one or more of the saxophone players can double on the clarinet, flute, or even oboe, and in performance these softer wind instruments may require amplification. Brass doublings may include the fluegel horn, French horn, and bass trombone. The pianist may double on one or more keyboard-based synthesizers.

Additional players may be required in the percussion section, and of course vocalists, with amplification, are common.

There are few formal traditions in seating of the big band, but generally the saxophones are in front, with the trombones on risers behind them. In many cases, the trumpet players stand behind the trombones, but they may also be seated behind them on a higher set of risers. The remaining players are to one side, and the rhythm group (piano, bass, and drums) are normally in proximity in order to maintain good ensemble.

Brass and saxophone players normally stand when they are playing solos; amplification may or may not be required with the saxophones.

Modern rock ensembles may number no more than five or six players. With the exception of drums and vocal parts, they are basically electronic, and of course can play at extremely high levels. The larger the performance venue, the more amplifiers and loudspeakers are involved.

There are no formal ensembles as such, and each group evolves out of the particular talents at hand. Generally, there are at least three guitars: lead guitar, rhythm guitar, and bass guitar. As a rule, most of the performers can sing, as harmonic backup to a lead vocal. A typical large group might consist of drums, electric bass, lead guitar, rhythm guitar, electric keyboards, lead vocal, and background vocal. Instrumentalists often double in the background vocal function.

10.7   THE MUSICAL THEATER

As in the opera house, the orchestra in musical theater is located in a pit. Normally, performances take place in theaters that are somewhat smaller than opera houses (often legitimate drama theaters), and there is usually limited seating space in the pit. Therefore, the "pit orchestra" is considerably abbreviated in its resources, as shown below.

5 1st Violins	1 Flute	2 Trombones	1 Percussion
4 2nd Violins	1 Oboe	1 Drum set	1 Harp
2 Violas	1 Clarinet		1 Keyboard
2 Cellos	1 Bassoon		1 Guitar
1 Bass	2 Trumpets		

Arrangements are normally made specifically for a given group, and doublings are common in all but the string section. The winds may double on saxophones as well as on other members of the flute, clarinet, and double reed groups. Rarely, some brass players may double on horn. The keyboard instruments include piano and celesta. The percussionist may double on timpani.

It can easily be seen that the pit orchestra tries to be all things to all arrangers, inasmuch as it can fulfill most jazz needs as well as conventional orchestral accompaniments.

10.8   THE SYMPHONIC BAND

The symphonic band is largely a product of the American secondary school system, and in many parts of the United States it is the chief instrumental ensemble. The normal instrumentation is given below.

2 Piccolos	24 B-flat clarinets	1 Tenor saxophone	2 Baritone horns
6 Flutes	2 Alto clarinets	1 Baritone saxophone	4 Trombones
2 E-flat clarinets	2 Bass clarinets	6 Cornets	5 Tubas
4 Oboes	4 Bassoons	4 Trumpets	6 Percussion
1 English horn	2 Alto saxophones	8 Horns	

The large number of clarinets is a consequence of their filling the role normally allocated to violins and violas in the orchestra. They are normally divided into three groups of eight players.

The symphonic band evolved out of the military band, and over the years a substantial body of original composition for the ensemble has come into being. In addition, arrangements and transcriptions abound, and most concert performances by bands feature these prominently. The symphonic wind ensemble is composed of the same instruments, but it is normally a smaller group than the symphonic band. The symphonic band requires careful balancing in order to do justice to its repertory. It is all too easy for it to play loudly much of the time, because of its extensive brass and wind resources. A true *pianissimo* is difficult, but when it is attained, it is a unique texture.

10.9   REQUIREMENTS FOR GOOD ENSEMBLE AND INTONATION

The chief requirement for control of pitch and rhythm is that musicians be able to hear one another. While concert halls may be designed primarily for the audience, considerable attention is often paid to details of the orchestra shell for the sake of the players. A tradeoff is required between reflecting sound outward and keeping some of the sound on stage where it will benefit the players. Another tradeoff is required in allocating space per player. In a symphony orchestra, the average space occupied per player is about 2 m^2 (21.5 ft^2). The result is that opposite sides of the orchestra can be quite far apart. In a chorus, the space requirement is about half this, and singers appreciate the immediacy of others around them.

Musicians generally favor reverberant spaces over those that tend to the dry side. In particular, even small volunteer choruses can sound quite professional when performing in live houses of worship.

Under the right conditions, and with only a little practice, many instrumental groups can perform without a conductor. Through most of the Classical period, the orchestra was "led" by the concertmaster, who gave little more than an occasional downbeat. The art of the conductor, and his subsequent rise to superstardom, is essentially a product of the last 150 years.

Another requirement for good ensemble is a noise-free environment. Such sounds as noisy lighting dimmers, air handling, and the like, can be very disturbing to musicians. Musicians are generally concerned with creature comforts, and every effort should be made to satisfy them.

10.10   SOUND PRESSURE LEVELS WITHIN ENSEMBLES

Listeners in a concert hall normally hear music in the range of 70 to 90 dB-SPL. Accordingly, sound pressure levels within the ensembles themselves may be quite a bit higher than this. Woolford (1984) describes peak levels at a distance of 1 m from the bells of brass instruments as being in the neighborhood of 130 dB-SPL! Benade (1972, 251) points out that levels at the left ear of a violinist, and at the right ear of a flutist, may be in the 90- to 100-dB-SPL range, while a loudly played piccolo can reach the player's threshold of pain.

Organ consoles are often located quite near the pipework, and levels well over 100 dB-SPL can be generated at the player's position.

The choices do not always rest with the individual players, and in many symphony orchestras, transparent plastic baffles are used to shield string players from nearby trumpets and trombones. There have been grievances registered with orchestra managements, and Woolford points out that, over a number of years, some orchestral players have in fact suffered some hearing loss that can be directly associated with their work.

As a general rule, ensembles rarely work more than 6 hours in a given day, although individual practice may prolong the exposure to high levels. A review of the OSHA and EPA criteria (see section 2.2.2) will show that the music profession can indeed be hazardous to one's hearing.

Within choral ensembles, long-term average levels are generally in the 80- to 100-dB-SPL range, so there is cause for concern here as well (Ternstrom and Sundberg, 1988).

10.11   SPECTRA OF ENSEMBLES

The orchestral spectrum, as shown in figure 10-11, indicates a maximum in the 500-Hz range, with overall response rolling off below as well as above this point. These data relate quite naturally to the spectral characteristics of the individual instruments, as shown in chapters 5 through 9. The rolloff above 500 Hz is very nearly 10 dB per octave.

These observations are consistent with Veneklasen's measurements, which are shown in figure 10-5. Note here that the string ensemble produces a smoother overall spectrum than do the winds and brass.

Outside of the "singers formant," which is primarily used by soloists, the spectrum of a chorus does not differ markedly from that of an orchestra.

Some modern ensembles, notably those that are amplified, produce fairly smooth spectra out as far as 10 kHz. This is largely the result of amplification of percussive sounds picked up at close quarters, as well as some spectral equalization in the electronics chain.

REFERENCES

Benade, A. 1972. *Fundamentals of Musical Acoustics.* London: Oxford University Press.

Beranek, L. 1962. *Music, Acoustics & Architecture.* New York: Wiley.

Culshaw, J. 1957. *Ring Resounding.* New York: Viking Press.

Eargle, J. 1986. *Handbook of Recording Engineering.* New York: Van Nostrand Reinhold.

Forsyth, M. 1985. *Buildings for Music.* Cambridge, MA: MIT Press.

Meyer, J. 1978. *Acoustics and the Performance of Music.* Translated by Bowsher and Westphal. Frankfurt: Verlag des Musikinstrument.

Ranada, D. 1986. "The Orchestral Image." *Opus Magazine* 2(6).

Sundberg, J. 1982. "Perception of Singing." In *The Psychology of Music,* edited by D. Deutsch. New York: Academic Press.

Ternstrom, S., and J. Sundberg. 1988. "Intonation of Choir Singers." *J. Acoustical Society of America* 84(1).

Veneklasen, P. 1986. "Science in the Service of the Performing Arts." Santa Monica, CA: Paul S. Veneklasen Research Foundation.

Woolford, D. 1984. "Sound Pressure Levels in Symphony Orchestras and Hearing." Paper presented at Australian Regional Audio Engineering Society Convention, 25-27 September, Melbourne. (AES preprint number 2104.)

11

Music and Speech in Performance Environments

In this chapter, we will extend the notion of sound fields, which were introduced in chapter 1. We will expand the concept of the reverberant field to include specific early reflections, as well as the later "statistical" reflections, in both large and small performance spaces. The specific goals of these spaces will be discussed, as will problems and compromises that acoustical designers have grappled with over the years. Problems of noise and acoustical isolation will also be discussed.

While there are no simple rules that ensure success in designing large performance spaces, there are certain principles that musicians and acousticians alike would agree on. Problems, when they are encountered, are usually the result of economic imperatives coming up against esthetic ones.

This lengthy chapter begins with a discussion of acoustics of small enclosed spaces and the nature of discrete modes, then moves on to large spaces. Requirements for musical performance in large spaces are discussed, followed by subjective aspects and performance measurements. The problems and opportunities for electronic enhancement of music performance are then presented. Halls for speech, small-scale music performance, and worship activities are analyzed, and, finally, the nature of acoustical isolation and noise control are discussed.

11.1 ACOUSTICS OF ENCLOSED SPACES

Imagine a very small room, perhaps no more than a meter or two on a side. Let the walls be quite rigid, and let us place a small loudspeaker in one wall of the room. Let us drive the loudspeaker at very low frequencies and analyze what is going on in the room.

The first thing we would observe is that the sound pressure is essentially uniform throughout the room. If we measured it carefully, we would find that it is proportional to the change in the volume of air in the room caused by the displacement of the loudspeaker's cone.

We can change the driving frequency of the loudspeaker over a substantial range, and, as long as the cone's displacement remains constant, the pressure in the space will remain as before.

If we continue to raise the driving frequency, we will eventually reach a frequency at which we observe an uneven distribution of pressure in the space. The first such frequency will occur when the longest dimension of the room is equal to one-half wavelength at the driving frequency.

For example, assume that the longest dimension is 2 m and that the loudspeaker is located at one end of that dimension. The frequency would then be that which has a wavelength of 4 m, or 86 Hz, and we would measure a very strong buildup of pressure at the walls bounding the longest dimension, while at the midpoint in the room we would detect an antinode, or virtually no pressure at all. Obviously, the room is in resonance, just like the various cylindrical and conical resonators we studied in earlier chapters.

As we progress upward in frequency, we will observe that the characteristic resonance frequencies, or normal modes, of the space become closer together. We will also note that many of them are harmonically related, but that there are other frequencies that are not. For a simple rectangular space, the normal modal frequencies f of the space are given by the following equation:

$$f = \frac{c}{2} \sqrt{ \left(\frac{n_l}{l} \right)^2 + \left(\frac{n_w}{w} \right)^2 + \left(\frac{n_h}{h} \right)^2 } \qquad (11.1)$$

where c is the velocity of sound in air; l, w, and h are the dimensions of the space; and n_l, n_w, and n_h are, respectively, integers that are independent of one another. The equation is independent of units of length.

In the case we observed before, the lowest mode in the 2-m direction of our experimental room would be given by setting n_l equal to 1, while n_w and n_h are set to 0.

By way of terminology, those modes that involve only a single dimension are called *axial*, those involving two dimensions are called *tangential*, and those involving all three dimensions are called *oblique*. Along a given room dimension, the set of axial modes will be harmonically related.

The mode structure in a rectangular space is relatively easy to analyze, as is that in other simple volumes, such as the sphere. However, complex volumes are best studied by what is called finite element analysis, a technique involving extensive computer calculations.

11.1.1   Density of Modes

Figure 11-1 shows the measured mode structure in a rectangular space with dimensions of 5.2 by 6.4 by 2.7 m. The trio of numbers above each response peak gives the pertinent values of n_l, n_w, and n_h. Note that modal density increases with frequency. As a convenient rule of thumb, we can say that at frequencies above about ten times the lowest modal frequency, the mode structure is fairly uniform, allowing the space to be analyzed by statistical methods.

A better measure here is given by equation 1.16, which sets the frequency above which the modal density in a room is uniform and the space considered to produce a diffuse reverberant field. In the example shown in figure 11-1, the lowest mode is observed at about 27 Hz. At ten times this frequency, 270 Hz, we would expect the mode structure to be fairly dense. Assuming

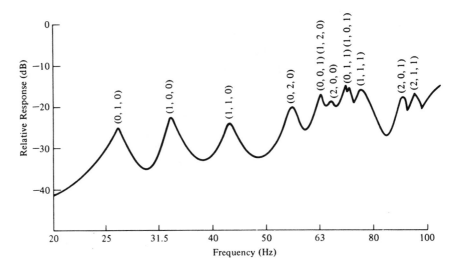

Figure 11-1. Mode structure in rectangular room.

that the space has a reverberation time of 1.5 sec, we can use equation 1.16 and arrive at the frequency of 258 Hz. Thus, our "ten-times" rule of thumb appears to be a good estimate for fairly reverberant spaces.

11.1.2 Large and Small Rooms

Consider two distinctly different spaces: a concert hall and the marbled vestibule of a typical public building erected half a century ago. Both spaces may have the same measured reverberation time at mid-frequencies, but a blindfolded listener will have no difficulty telling them apart immediately.

For the concert hall, the reverberation time is given by

$$T = \frac{0.16V_1}{S_1\overline{\alpha}_1}$$

while that of the vestibule is given by

$$T = \frac{0.16V_2}{S_2\overline{\alpha}_2}$$

Since the two values of reverberation time are equal, we can write

$$\frac{0.16V_1}{S_1\overline{\alpha}_1} = \frac{0.16V_2}{S_2\overline{\alpha}_2}$$

Now, let us assume that the volume of the concert hall is about nine times that of the vestibule. Then we can write

$$\frac{0.16V_1}{S_1\bar{\alpha}_1} = \frac{0.16(9V_1)}{S_2\bar{\alpha}_2}$$

which reduces to

$$S_1\bar{\alpha}_1 = \frac{S_2\bar{\alpha}_2}{9}$$

Recalling equation 1.17, we can now determine critical distance in the two spaces for an omnidirectional radiator.

For the concert hall,

$$D_C = 0.14 \sqrt{S_1\bar{\alpha}_1}$$

For the vestibule,

$$D_C = \left(\frac{0.14}{3}\right) \sqrt{(S_1\bar{\alpha}_1}$$

We see that the critical distance in the vestibule is only one-third as great as that in the concert hall, which means that the direct-to-reverberant ratio in the vestibule will be much less than in the concert hall. A listener at a given distance from an omnidirectional sound source in the vestibule would hear reverberant sounds about 10 dB greater than at the same distance in the concert hall [since 10 log (1/3) = 10]. For this reason, speech communication over relatively small distances will be more difficult in the vestibule than in the concert hall.

Another significant difference between the two spaces is the mean free path (MFP) that sound travels before it encounters an obstacle

$$MFP = \frac{4V}{S} \tag{11.2}$$

Let us assume that the concert hall and the vestibule have the following volumes and surface areas:

$$V_1 = 18{,}000 \text{ m}^3$$

$$S_1 = 4260 \text{ m}^2$$

$$V_2 = 2000 \text{ m}^3$$

$$S_2 = 985 \text{ m}^2$$

Then, for the concert hall,

$$MFP = 17 \text{ m}$$

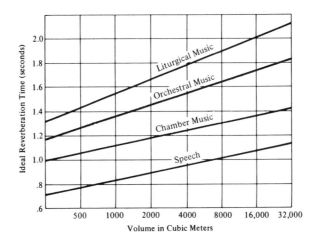

Figure 11-2. Suggested reverberation time versus volume for various activities.

and for the vestibule,

$$MFP = 8 \text{ m}$$

What this implies is that the initial time delay (ITD) between early reflections in the two spaces will be about twice as long in the concert hall as in the vestibule. The implication is that a blindfolded listener would be able to estimate the physical extent of each space, at least in relative terms, purely from the lateral reflections of impulsive sounds.

Another difference between the two spaces that might be perceptible is the difference in modal density at low frequencies. Assuming that the reverberation time in both spaces is 2.5 sec, we can again use equation 1.16 to determine the Schroeder frequency (see section 1.7.5). For the concert hall, this is

$$f = 2000 \sqrt{\frac{2.5}{18,000}} = 23 \text{ Hz}$$

and for the vestibule,

$$f = 2000 \sqrt{\frac{2.5}{2000}} = 71 \text{ Hz}$$

The conclusion we draw here is that the listener can identify the size of a space, at least partially, through the density of low-frequency room modes, most notably by the gaps that may exist between them.

Thus, the reverberation time of a space is only one of many elements to be considered in assessing its performance. In general, the smaller a space is, the lower its reverberation time normally should be, primarily in order to keep direct-to-reverberant ratios high enough for clarity of both speech and music. Figure 11-2 gives an indication of the general practice in this area.

11.1.3   Statistical Versus Real-World Conditions

Thus far, we have assumed that diffusion is uniform above the Schroeder frequency, and that reverberation-time equations describe absolutely smooth decay curves. In the real world, things are not that orderly. If absorption in the space is uniformly distributed over the entire area, then we have a good chance of observing the reverberation times as given by the Sabine equation (equation 1.15).

Schroeder (1984) points out how profoundly the distribution of absorption in a space can change the measured reverberation time in that space, as shown in figure 11-3. In one case (figure 11-3a), the totally absorbing surface is located on one wall, and it is exposed to maximum incident acoustical power. Thus, it is effective as an absorber, and the measured reverberation time is low. In the second case (figure 11-3b), the same amount of absorptive material is located in a corner, and thus is not exposed to as much incident acoustical power as in the previous case. One may say that the two sections of absorption largely "see each other," rather than the room, and so their overall absorptive effect is lessened, resulting in a longer reverberation time.

In other cases of nonuniform absorption, we often encounter rooms in which the floor and ceiling may be covered with fairly absorptive materials, while one or more opposite pairs of walls may be relatively reflective. In these cases we may observe a reverberation decay curve with two separate slopes, one showing rapid decay and the other showing a slower decay. As shown in figure 11-4, the rapid decay occurs primarily in the room modes involving the surfaces with higher absorption, while the slower decay rate occurs in the room modes involving the less absorptive surfaces. In such rooms we will most assuredly observe "flutter echoes" between the wall pairs with lesser absorption. Such reflections take place in any space, but individual echoes are more noticeable here because of the overall lower density of reflections.

None of the standard reverberation-time equations will give us any indication of the double-sloped decay curve; however, a variant of the reverberation-time equation developed by Fitzroy (1959) will give an average of decay rates along the three dimensions of a rectangular space.

Obviously, acoustical designers take reverberation-time calculations as only broad estimates, noting carefully all assumptions that may have been made. Reverberation-time measurements are of course real-world events, and in the case of multiple slopes most acousticians consider the initial decay rate as the most significant.

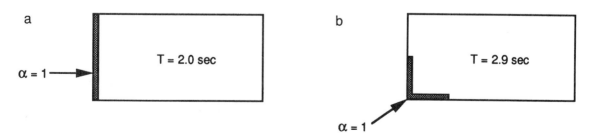

Figure 11-3. Dependence of reverberation time on the distribution of absorption in a space. Data after Schroeder 1984.

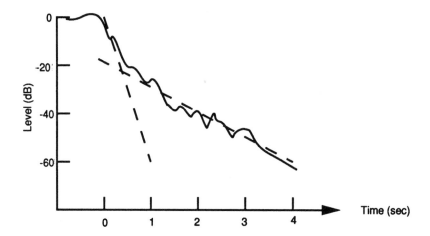

Figure 11-4. A double reverberant decay curve in a space with uneven distribution of absorption.

11.2 REQUIREMENTS FOR MUSIC PERFORMANCE SPACES

The vast majority of musicians, acousticians, and music listeners would agree that music should be presented against a texture of overall spaciousness; however, that spaciousness should not be so pervasive as to blur the music or otherwise get in its way. All good performance environments, regardless of their size, support this idea in one way or another.

The key factor here is discrete lateral reflections following the onset of direct sound within a short interval. Reverberation time and level per se are of lesser significance.

Regarding early reflections and the sensation of spaciousness, Kuttruff (1979) states:

1. The reflected sound signals must be mutually incoherent.

2. Their intensities must be above a certain threshold.

3. Their delay with respect to the direct sound must not exceed 100 msec.

4. They must arrive from lateral directions.

The requirement for mutual incoherence basically states that identical sets of reflections should not arrive at the listener from both sides of the room. This condition is normally met through departures from symmetry in the location of both musicians and listeners.

Regarding threshold, the reflections must be perceptible to the listener. For a delay of 50 msec, a signal arriving at a horizontal angle in the range of 45° will be perceptible to a listener even when it is 20 to 25 dB lower relative to the direct sound arriving from straight ahead. The threshold will vary for other values of delay and arrival angle.

Barron (1971) defines the subjective relationships between delay times and direct-reflected ratios for a reflected sound incident at 40°, as shown in figure 11-5. Note that reflected sound should be no lower than 20 to 25 dB, relative to direct sound, if it is to contribute to an impression of spaciousness.

Delayed signals arriving via reflections directly overhead do not normally contribute to the sense of spaciousness; rather, they are perceived as strengthening the direct signal and may be useful toward the rear of the house in maintaining the loudness of stage events. However, the acoustician must keep in mind that reflected sound from the stage will ultimately satisfy more patrons if it reaches them via lateral reflections rather than as overhead reflections.

11.2.1   The Acoustical Designer's Task

The reader can appreciate the difficulties facing the acoustical designer of a concert hall. First, there is the architect's conception of the project, which he may think of purely in visual and spatial terms. Many of the things that the acoustician may wish to do to promote good acoustics will be summarily dismissed by the architect because they conflict with his master plan. In short, the architect is often loathe to copy a tried and true architectural design that the acoustician knows will work well.

Then there are the requirements of music management. In order for the hall to pay for itself, it must be able to accommodate a certain number of patrons. That number will probably be larger than the acoustician would consider optimum for proper acoustics. In addition, management may insist that the hall be adaptable for opera, thus necessitating that there be a proscenium with a removable orchestra shell. If this is the case, the acoustician may simply give up any notion of designing an ideal environment for symphonic performance. At this point, he must realize that the results will be mixed and that he runs a great risk of being damned by the critical press. Artful compromises are called for; let us examine them in the context of specific concert hall designs.

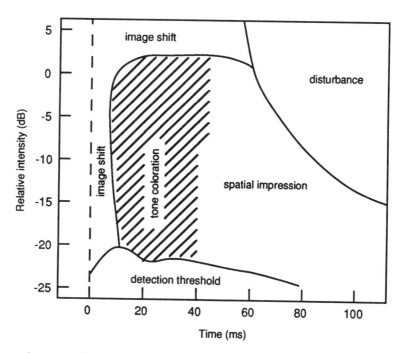

Figure 11-5. Subjective effects of delay and delay level for reflected sound incident from 40°. Data after Barron 1971.

The so-called shoebox design is typified by many older European halls and in the United States most notably by Symphony Hall in Boston. The general form for the structure is shown in figure 11-6. Such structures usually have coffered ceilings, and the walls usually have architectural relief details, both of which promote diffusion. The stage is shallow, and the proscenium, if there is one, is high, with angles set to reflect sound outward. If the seating in the hall is limited to no more than about 2500, and the volume to about 19,000 m³, then the time gap between the arrival of direct sound and the first reflections will be in the range of 15 msec, because of reflections from both the side walls and the balcony railings. This condition can be maintained over most of the main floor, but it is clear that the patrons in the balcony will not enjoy the same acoustics as those seated at orchestra level.

The acoustician is concerned with the balance between discrete side reflections and diffuse reflections. The sound must ultimately be diffuse, but strong early lateral reflections are to be desired. For a given room volume, reverberation time and direct-to-reverberant relationships will be more or less fixed relative to each other. The acoustician must target a reverberation time that satisfies these musical considerations in all parts of the house. It goes without saying that the total volume of the hall must be sufficient to support the intended reverberation time and direct-to-reverberant relationship.

From the point of view of the musicians, the stage area must provide reflections that enable them to hear one another comfortably. Such reflections normally come from the ceiling above them, and here there is a tradeoff between the amount of sound that is reflected downward toward the players and the amount that is reflected outward to the audience.

A common complaint in many concert halls is lack of bass. The problem is often related to too much low-frequency absorption and insufficient reverberation time at low frequencies. In some cases, the orchestra shell may not be rigid enough, with consequent flexing and sound absorption at low frequencies.

Many acousticians favor the relationship between low-, mid-, and high-frequency reverberation times shown in figure 11-7, especially in halls where large orchestral works of the nineteenth century will be performed. The curve shown in figure 11-7 is a natural consequence in many locations, since rigid structures offer little absorption at low frequencies, and excess absorption in air at high frequencies is always present.

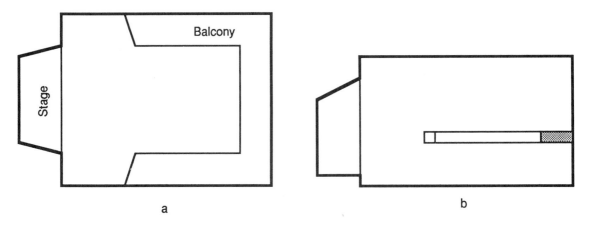

Figure 11-6. A typical "shoebox" concert hall. (a) Plan view; (b) elevation view.

Another culprit in the missing bass problem is simply the spacing between seating rows at orchestra level. Sound from the orchestra strikes the seating at a very low grazing angle of incidence, and the regular spacing of the rows provides a "trap" for sound in the 125-Hz range. Higher overall reverberation time may tend to swamp this effect out.

Balconies require special treatment. In halls intended for music, recesses under balconies should ideally have overhead clearance higher than they are deep, in order to allow some degree of lateral reflected sound to enter. Deep recesses are very unsatisfactory.

In an effort to accommodate larger audiences, many architects have been drawn to the fan-shaped structure shown in figure 11-8. This design accommodates more patrons with overall shorter sight lines, but the acoustical price paid is loss of control of lateral reflections. As can be seen in the figure, discrete reflections from the sides of the hall are reflected toward the back of the space, and most of the audience will miss them altogether. The sides of a fan-shaped hall may be articulated as shown in figure 11-9, so that useful reflections can be

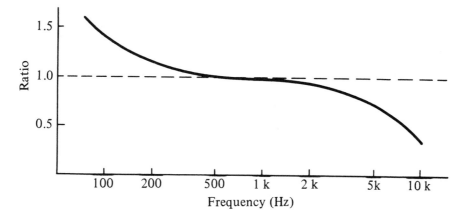

Figure 11-7. Normal variation of low- and high-frequency reverberation time in a concert hall relative to the mid-frequency value.

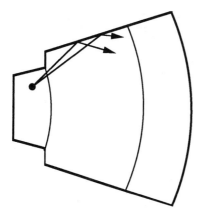

Figure 11-8. Plan view of a fan-shaped concert hall.

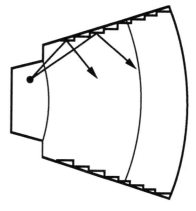

Figure 11-9. Plan view of a fan-shaped concert hall with articulated side walls.

generated. However, for those patrons in the middle of the house, the lateral reflections may be too late, creating disturbing echoes.

When the seating capacity is extended to 3000 persons or more, the volume of the space must be increased accordingly, and there are apt to be problems in simply getting enough loudness from the orchestra to satisfy most listeners. Some acousticians have made good use of "clouds" suspended overhead to control early reflections, as shown in figure 11-10. If the array of clouds is dense enough, then reflectivity from them can be fairly broad across the frequency band, as Beranek (1962) has shown. However, if they are widely separated, the clouds may reflect sounds in the range of 500 Hz and higher, promoting a sense of intimacy. At low frequencies, widely spaced clouds are effectively not present. In either case, the entire volume of the space contributes to the development of reverberation. Properly utilized, clouds can give some of the intimacy of a smaller space without the disadvantages of less volume.

The so-called horseshoe shape of many opera houses is traditional, and the many tiers make it possible to accommodate the maximum number of patrons with the shortest sight

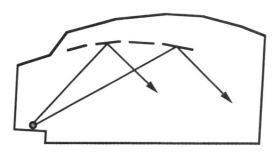

Figure 11-10. Section view of a concert hall with "clouds."

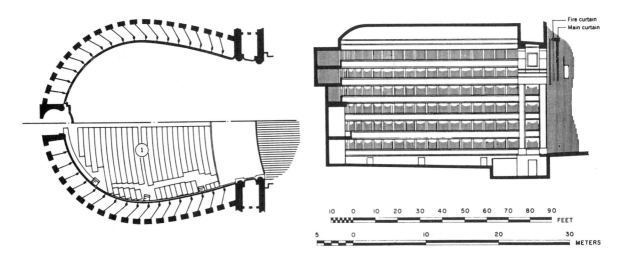

Figure 11-11. Views of a horseshoe-shaped opera house. From L. Beranek 1962, courtesy of J. Wiley and Sons, New York.

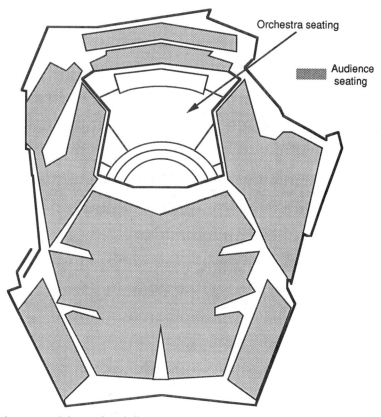

Orchestra seating

Audience
seating

Figure 11-12. Plan view of the Berlin Philharmonie.

lines to the stage, as shown in figure 11-11. Opera is seen as well as heard, and visual immediacy is highly valued. Early reflections from the many balcony railings are useful, but since so much of the space is absorptive, the reverberation time in most opera houses is quite short.

The orchestra is relegated to placement in a pit in front of the stage, and of course the singers perform on a highly absorptive stage. It is a difficult acoustical environment for all but the hardiest singers.

Innovative approaches in concert hall design have been more popular in Europe than in the United States. The reasons go back many years. Traditionally, European concert halls had no proscenium and were equipped with large organs and permanent choir seating behind the orchestra. This effectively jutted the orchestra out into the hall, and on occasions when the choir was not present, patrons were often seated to the sides of, and even behind, the orchestra. Many of the newer European concert hall designs have taken this as a point of departure, placing the orchestra even more toward the center of the room. Often, there is liberal use of overhead reflectors to achieve control of early reflections, but it is clear that patrons to the side of and behind the orchestra may hear unusual instrumental balances. Sight lines are usually excellent for all patrons, and there is often a feeling of immediacy and involvement with the performers. Figure 11-12 shows a typical nontraditional approach in which patron seating surrounds the curved tiers on which the performers sit.

11.2.2   Modeling and Computer Simulation of Halls

Today a major concert hall is rarely built without some degree of modeling. The usual method is to construct a one-tenth scale model and drive it with frequencies ten times normal. Thus, performance at 4 kHz will be modeled with a 40-kHz signal. The materials used in constructing the model are critical, since they must have a times-ten relationship in absorption coefficient with the materials to be used in the finishing of the actual hall. Air absorption in the model must also be carefully assessed and compensated for. Models are of course quite expensive to build, and much time will be taken in measurement, analysis, and subsequent modification.

On a much simpler scale, certain two-dimensional aspects of acoustical performance may be modeled by what is called a ripple tank. This is a shallow pool of water in which the effects of diffraction and reflection can be directly observed as transverse waves in the water.

In this age of rapid computation, many people are surprised to find that computer modeling of complex acoustical spaces has not been developed to a point of overall confidence. Certain aspects of spaces can be relatively easily modeled—specifically, specular reflection from plane surfaces. Such models can easily spot certain kinds of problems beforehand. Diffraction, on the other hand, is very complex, and there are no simple models. At present, given the amount of detail required for useful and complete modeling of a concert hall, plus the cost of computing, computer modeling will be equivalent in effort and monetary outlay to building a physical model. However, given the rapid development of computational hardware and methods, we may expect to see comprehensive computer models of concert halls in the reasonably near future.

As an example of specular modeling at high frequencies, figure 11-13 shows first-, second-, and third-order reflections from a source on stage to a given listening position in the house.

11.2.3   Subjective Aspects of Performance Spaces

The requirements for music of one period are not the same as those for another. Orchestral music of the Classical period was written for ensembles numbering no more than thirty-five or forty players, and performances took place in modest-size halls. The music took advantage of the characteristics of the performance space, and there is no question that it sounds best in such spaces.

Intimacy is a term often used to describe the proximity of the listener to the performers, and the initial time gap between direct sound and the first reflections is often no more than 15 or 20 msec. Today, Classical orchestral music of such composers as Haydn (1732–1809) and Mozart (1756–1791) sounds best when played in halls seating no more than about 1200 persons. While scaled-down ensembles are often used for Classical performances in larger halls, even those seating as many as 3000, the results are not always satisfactory because of insufficient loudness from the ensemble. Any kind of chamber music or piano music usually sounds best in smaller halls for the same reasons.

The term *liveness* refers to a relatively high amount of reverberation in the mid-frequency band (0.5 to 2 kHz), along with fairly short initial time gaps. Many older halls exhibited this characteristic, and it is especially desirable for vocal and instrumental solo writing. Performers enjoy such an environment because it often provides them immediate feedback from the hall. Many small recital halls have been designed with this in mind.

The term *warmth* characterizes larger halls that are noted for performance of large nine-

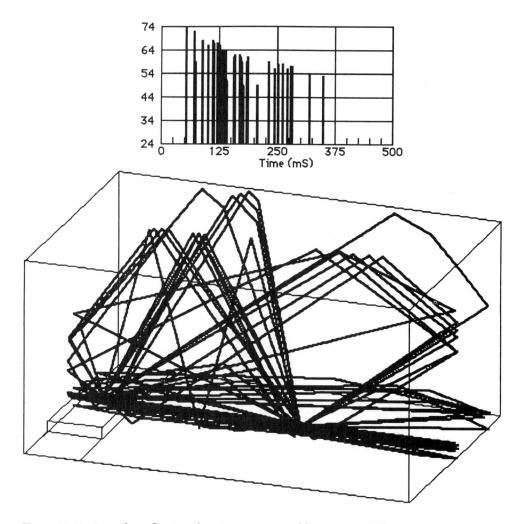

Figure 11-13. Specular reflections between source and listener in a hall.

teenth-century and modern orchestral works. It is characterized by a rise in reverberation time at low frequencies (below 250 Hz) relative to higher frequencies. Relatively large volume is necessary to accomplish this, and accordingly, small groups or solo instruments will normally not fare well in such spaces.

Obviously, one hall cannot satisfy all the musical demands that may be made on it, and again the acoustician is challenged to make the best compromises.

11.2.4 Measures of Performance

For years acousticians have wanted to develop methods of quantifying the performance of concert halls, attaching numbers to such concepts as clarity, warmth, and spaciousness.

Schultz (1965) proposed a measure of "running reverberance" as an indicator of perceived liveness in a hall. Mathematically, it is defined as

$$R = 10 \log \left\{ \frac{\int_{50\,ms}^{400\,ms} p^2(t)\,dt}{\int_{0\,ms}^{50\,ms} p^2(t)\,dt} \right\} \quad dB$$

Here, p^2 represents the square of sound pressure, which is proportional to acoustical power.

Essentially, this is a measure of reverberant-to-direct sound in the hall. Figure 11-14 shows measurements in three concert halls. If the value is too great, then clarity suffers and musical definition is degraded.

Alim (1974, 243) proposed the measure of "clarity index," which is defined as follows:

$$C_{80} = 10 \log \left\{ \frac{\int_0^{80\,ms} [g(t)]^2\,dt}{\int_{80\,ms}^{\infty} [g(t)]^2\,dt} \right\} \quad dB$$

where g represents a weighted pressure function.

A value of C_{80} of 0 dB is considered excellent, affording clarity to complex musical passages. A value of -3 dB can be tolerated in most situations.

Note that C_{80} and Schultz's R are expressed in decibels; one measurement is effectively the inverse of the other. Positive values of R correspond roughly to negative C_{80} values.

11.3   ELECTRONIC CONCERT HALLS

The problems inherent in multipurpose halls have raised the possibility of electronic enhancement as an alternative to traditional acoustics. Since the early 1970s, digital delay devices have been available with levels of performance and reliability that enable designers to specify a wide range of electronic enhancement schemes, some of which have produced excellent results. The major problem remains that of making a small room seem larger, or acoustically more live than it naturally is; considerable skill and taste are required to do this well. In this section we will examine some of the techniques that have been used for these purposes.

11.3.1   Assisted Resonance

This enhancement method was developed by Parkin et al. (1975, 169–179) and was used in Royal Festival Hall in London. When that hall was opened in the early 1950s, many patrons and critics felt that it lacked warmth. There were no acoustical options available for increasing reverberation time at lower frequencies, so the idea of assisted resonance was developed. Details are shown in figure 11-15.

There is an assemblage of microphone–amplifier–loudspeaker channels, each with the microphone enclosed in a Helmholtz resonator. Approximately 170 channels are used, and

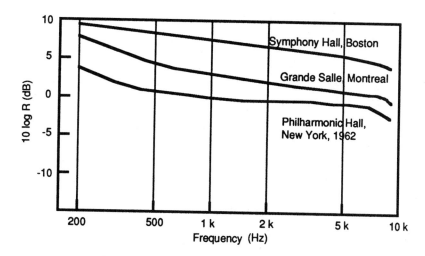

Figure 11-14. Running reverberation plots in three halls.

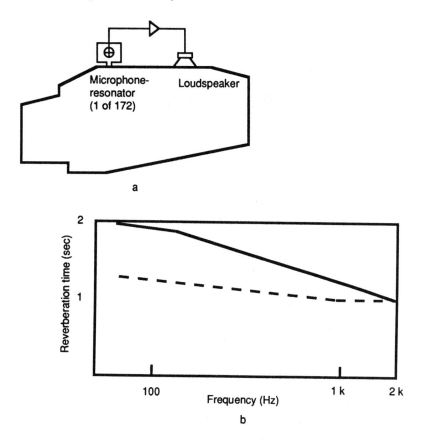

Figure 11-15. Assisted resonance system, (a) Schematic; (b) dashed line—system off; solid line—system on.

each channel responds only to the tuning of its resonator. The frequency range covered by the assemblage of channels is from 60 to 700 Hz, and the devices are located in the ceiling of the hall. Note that the reverberation time can be increased by about a factor of 2 at the lowest frequencies.

The assisted resonance approach addresses only the lower frequencies. In the case of Royal Festival Hall, the pattern of early lateral reflections was deemed to be satisfactory in most respects, requiring no remedial work.

11.3.2   Stage-to-Hall Coupling

Paul Veneklasen (1975) noted that the reverberant character of many large stage houses could be put to good use by coupling that sound field electrically into the hall itself, as shown in figure 11-16. Several caveats are in order here. First, the stage house must be very quiet so that backstage noise does not intrude. Depending on production requirements, the stage house may or may not be a quiet place. Also, depending on production requirements, the liveness of the stage house may vary considerably. In large halls designed for both opera and symphonic performance, the system can work well for symphonic music.

11.3.3   Sound Field Synthesis

This category of enhancement covers a variety of approaches, all making use of digital delay and reverberation devices. The general approach is to sample the sound field in the area over the performers and feed the stereophonic signals to a network of delays and reverberation simulation. One way in which this can be done is shown in figure 11-17. Discrete delays can simulate early reflections, and the overall reverberation time setting is modeled after some larger acoustical space. Through careful adjustment of all parameters, a convincing impression of large room size can be created. Veneklasen was the pioneer in this area, and later development has been carried out by Jaffe.

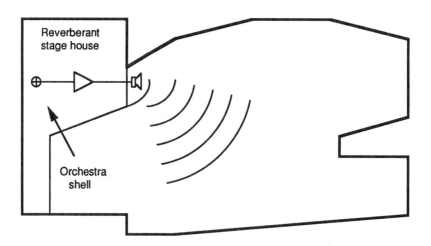

Figure 11-16. Stage-to-hall coupling.

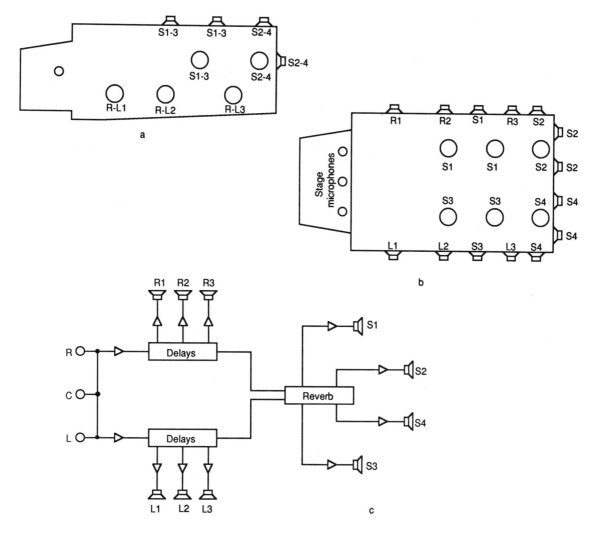

Figure 11-17. Sound field synthesis. (*a*) Section view; (*b*) plan view; (*c*) signal flow.

11.3.4   Sound Field Amplification

This enhancement method, developed by Philips of Holland (1980), consists of a large number (often more than 100) of microphone–amplifier–loudspeaker channels arrayed randomly around an auditorium. Each electroacoustical loop is operated at low level so that stability is ensured. Through careful adjustment, the reverberant-field level in the space can be effectively doubled (raised 3 dB), and the reverberation time itself doubled as well.

While sound field synthesis can make a room seem larger than it actually is, sound field amplification creates only the effect of increased reverberation time, with no apparent increase in room size.

11.3.5   General Comments

The traditional approach to concert hall acoustics has shunned the various options presented in this section, except in the case of Royal Festival Hall, where assisted resonance was brought in as the only possible remedy. The traditionalists may be right, at least in terms of what has been practiced thus far. Most, but by no means all, electronic enhancement schemes have been inadequately specified, and operation has been far from foolproof. Component reliability is a problem because of the complexity of the systems. A given component may have excellent statistical reliability, but an ensemble of them may not. It is one thing for an electronic component to simply stop working, and another thing for it to malfunction and produce noise or distortion. The latter would be absolutely intolerable in a modern concert hall and would bring immediate critical damnation.

Perhaps what is needed at the present state of development is a combination of good acoustical design and a modicum of electronic enhancement only in those remote parts of the hall where it may be needed. Also, the need for subtle early lateral reflections alone may be met electroacoustically in halls that may not support that need naturally.

11.4   HALLS FOR SPEECH

Today, most speech presentation is amplified; however, there was a day when this was not possible. As a starting point for our discussion, let us go back to early theaters and lecture rooms.

The paramount requirement is, of course, that speech be intelligible to the majority of the audience. Broadly, we can define the requirements:

1. *Loudness.* The speech signal must be loud enough that listeners do not have to strain to hear it. This requirement alone may limit the size of the room, and in larger theaters actors will need to project well and consistently throughout a performance.

2. *Speech-to-noise ratio.* When speech is presented at moderate levels, background noise levels must be 25 to 30 dB below speech peaks.

3. *Reverberation time.* The normal rate of speech articulation is three to five syllables per second. If the reverberation time is too long, then the decay of one syllable may not be quick enough to "get out of the way" of the next one. In most cases, reverberation time should be limited to values not exceeding about 1.2 sec.

4. *Direct-to-reverberant ratio.* This item is related to the previous one, in that longer values of reverberation time are directly related to diminution of the direct-to-reverberant ratio. Excessive reverberant level, regardless of actual reverberation time, may be perceived as noise behind the speech signal.

Lateral reflections should follow very shortly after the direct sound, and pronounced echoes are to be avoided at all costs. In lecture rooms, raked seating is often used because it both provides good sight lines and avoids the acoustical losses associated with a low grazing angle of incidence. Studies here go back to the work of Russell and his "isacoustic" curve, dating from 1838 (Forsyth 1985), as shown in figure 11-18.

Reflecting canopies overhead focus speech sound where it is needed, and side walls in the

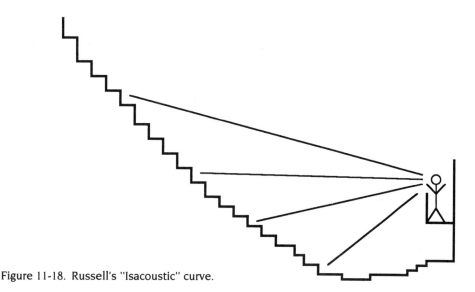

Figure 11-18. Russell's "Isacoustic" curve.

vicinity of the speaker should be angled to project sound outward. The average speech level of a typical talker is no more than about 65 dB-SPL at a distance of 1 m, and this corresponds to an acoustical power output of about 10 μW. No acoustical power can afford to be lost. Obviously, the requirements for speech intelligibility are different from those of music presentation. Lateral reflections may be useful, but the bulk of the energy should ideally arrive from the direction of the talker, and within a short time gap.

11.4.1   Theater Requirements

Figure 11-19 shows architectural details of a typical theater used for dramatic productions. It can be assumed that the stage is virtually nonreflecting and that the entire acoustical power in the house is the result of direct sound and reflections from the boundaries. The maximum front-to-back distance is about 15 m (50 ft), ensuring good involvement between audience and action. The short reverberation time of 1.2 sec supports the speech level of the actors, especially at the back of the house.

11.4.2   Measurements of Speech Intelligibility

The most straightforward way to determine speech intelligibility in a given space is to perform articulation tests. In such a test, a one-syllable word is embedded in a carrier sentence. Listeners in various parts of the house are asked to write down the word as they perceive it. Typical carrier sentences might be as follows:

"The first word to identify is *cat*."

"Now, I want you to write down the word *toy*."

The purpose of the carrier sentence is to place the word in the typical acoustical environment of flowing speech. The sentence indicates which word is to be written down, but it gives

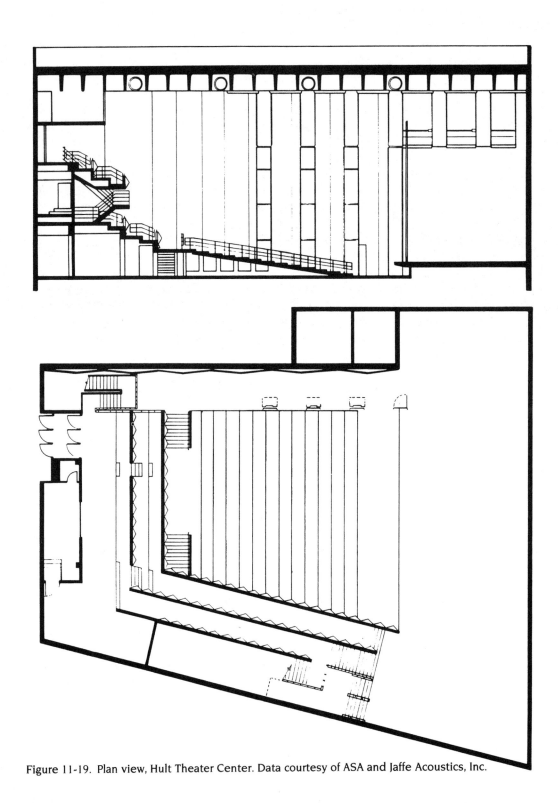

Figure 11-19. Plan view, Hult Theater Center. Data courtesy of ASA and Jaffe Acoustics, Inc.

no contextual indication of what that word is. A given test may consist of 100 trials.

Because of the redundant nature of language, a listener does not have to identify every word in order to understand the substance of what is being said. Generally, if a listener can identify 85% of the syllables in an articulation test correctly, then he will be able to understand normal speech in the same testing environment with an accuracy of 97%. Similarly, an intelligibility test score of 75% will indicate understanding of 94% of the words in normal speech context. Figure 11-20 shows the approximate relationship between intelligibility testing based on random syllables and the understanding of words in sentence context.

Going beyond direct measurements of speech intelligibility, French and Steinberg (1947, 90–119) developed a method of estimating the articulation index (AI) of telephone transmission lines. In such systems, reverberation was not a problem, and only the signal-to-noise ratio as a function of frequency was considered. Kryter (1962) simplified the calculation procedure, as shown in figure 11-21.

AI calculations are not normally used in acoustically live spaces such as theaters and lecture rooms, but the method is useful in estimating intelligibility of paging systems in noise-prone environments where reverberation is insignificant.

Peutz (1971) has developed a measure of the articulation loss of consonants that considers chiefly the effects of reverberation time and the direct-to-reverberant ratio at the listener. If the reverberation time in the octave band between 1 and 2 kHz is known or can be estimated, and if the signal-to-noise ratio is 25 dB or greater, then the articulation loss of consonants can be determined from the graph presented in figure 11-22. What is required is a simple calculation of the direct-to-reverberant ratio at some listening point in the house. The direct level is determined by inverse square relationships, and the reverberant level L_R is determined by

$$L_R = 126 + 10 \log (W_A) - 10 \log (S\overline{\alpha}) \qquad (11.3)$$

where W_A is the acoustical power from the sound source, and $S\overline{\alpha}$ is the total absorption in the space. $S\overline{\alpha}$ can be determined from the reverberation time equation:

$$S\overline{\alpha} = \frac{0.16V}{T} \qquad (11.4)$$

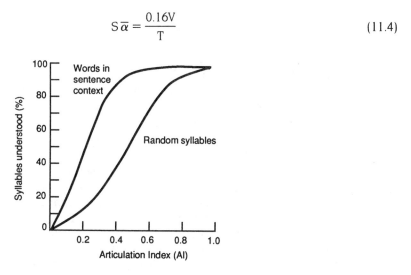

Figure 11-20. Correlation between intelligibility of random syllables and words in sentence context.

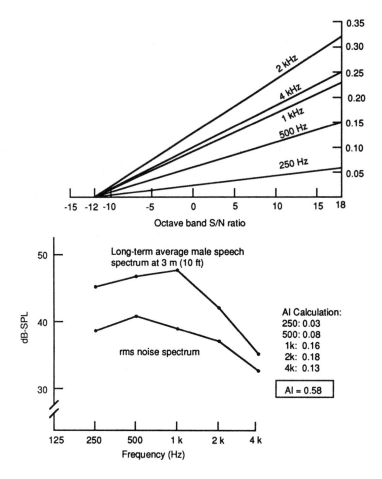

Figure 11-21. Calculation of articulation index (AI).

Houtgast and Steeneken (1972) developed the notion of modulation transfer function, noting that the loss of speech transmission index (STI) could be related to the erosion of the modulation index of a number of speech carrier frequencies at a number of modulation rates. Both noise and reverberation effects contribute to the diminution of modulation index.

A significant outgrowth of their work has been the development of a meter that directly measures the loss. Rapid speech transmission index (RASTI) is the name, and RASTI meters are now being used to determine in a minimum amount of time the performance of speakers in speech environments, with or without sound reinforcement. The signal source, a small loudspeaker, is placed at the position of the talker, and the analysis part of the apparatus is located at one of a number of listening positions. The RASTI signal output contains a set of modulation frequencies of noise bands centered at 500 and 2000 Hz. The major caveat is that there be no unusual low-frequency disturbances in the space to be measured.

Related to the notion of C_{80} discussed in section 11.2.4 is the clarity index, or *Deutlichkeit*, as proposed by Thiele (1953). The notation here is D, and the definition is

$$D = \frac{\int_0^{50 \text{ ms}} |g(t)|^2 \, dt}{\int_0^{\infty} |g(t)|^2 \, dt} \times 100\%$$

where $g(t)$ is related to the impulse response in the room. Again, we are looking at a measure of the direct-to-reverberant relationships of the room. Values of D generally correspond to speech intelligibility, as shown in Figure 11-23.

Much progress has been made during the decade of the 1980s in correlating measurement

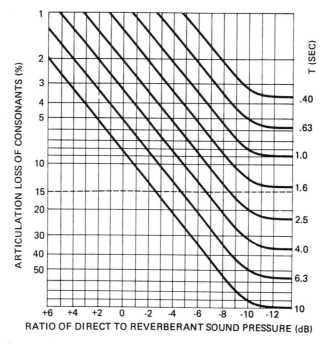

Figure 11-22. Calculation of Peutz' articulation loss of consonants.

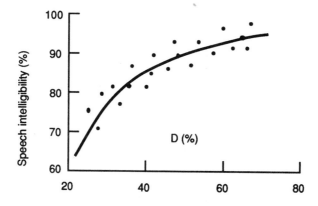

Figure 11-23. Correlation between D and speech intelligibility.

and estimation techniques for speech intelligibility. What has been learned from this is what to avoid at the specification and construction stages so that expensive redesign and reworking can be avoided. The public expectations for speech intelligibility, with or without sound reinforcement, are high, and mistakes are not often overlooked or tolerated.

11.5   SMALL HALLS FOR MUSIC PERFORMANCE

The same observations that apply to large performance spaces apply to smaller ones. There is one fundamental rule, and it is that there should be no attempt to design a small hall with the reverberation characteristics of a large one. To do so would be to create a hall in which the direct-to-reverberant ratio would be relatively low, with consequent blurring of musical textures. See the discussion in section 11.1.2 if this point does not seem clear.

The main use for a small hall is for performance of smaller musical forms, such as chamber ensembles and solo pieces. Ideally, the space should be finished with the same kinds of materials used in a larger concert hall. Intimacy, as we have earlier defined it, is desirable in small halls and comes almost as a byproduct of the size and dimensions of the space.

Many colleges have small halls, seating perhaps 700 persons, dedicated to pipe organ music. It is true that an organ can be installed in a smaller, quite live room and sound to advantage with up to 2 to 2.5 sec of reverberation time. The organ can in fact be voiced so that it will sound to advantage in such a space. Choral performances of early music will also sound to advantage in such a space, but solo piano or string quartet performances may suffer from too much reverberation and a too small direct-to-reverberant ratio.

If a hall is intended for such purposes, it is best if it is designed with variable acoustics. Rotating sections in the walls and ceiling can be designed that expose more or less absorptive surfaces and that can render the room acoustically appropriate for a variety of musical forms. The design of such a hall is tricky and is best left in the hands of an experienced acoustical consultant.

11.6   ACOUSTICS OF REHEARSAL SPACES

In most music complexes, the main halls are busy with performances, dress rehearsals, and various maintenance and changeover activities. Lucky is the ensemble that has its own hall and can schedule all its rehearsals there.

In many cases, and on almost all college campuses, large ensembles must rehearse in spaces built for that purpose. The floor area must be capable of handling the required number of players or singers. Ceiling height is, of course, the main problem. If the ceiling is too high, the space is expensive to build and to air condition. If the ceiling is too low, then musicians on one side of the room will have difficulty hearing those on the other, and the space will seem oppressive.

If the space is too live, then the music will be too loud, giving the performers the wrong sense of what will actually happen in performance. If it is too dead, then the musicians will fatigue easily. All these details must be worked out by a competent acoustician and a concerned architect.

The practice room is the smallest acoustical unit we are likely to encounter. It normally has space for a single upright piano and perhaps two persons. Needless to say, it should be quite absorptive and reasonably well isolated from neighboring practice rooms.

11.7   WORSHIP SPACES

Traditionally, worship spaces have presented an acoustical conflict between the requirements for spoken communication and the demands of liturgical music. When early Christianity gained acceptance, it developed large worship spaces and worship styles that made good use of long reverberation time.

In time, the pipe organ and choir became the center of musical activities, and verbal communication was largely transformed into various stylized chants and intoned responses. A simple sermon was understood only by those persons close to the pulpit. The old churches of Europe demonstrate this today, and only through speech reinforcement is it possible for all worshipers to understand the spoken word. Harris (1950) describes the use of Helmholtz resonators in old Scandinavian churches to minimize the excessive overhang of certain frequencies. Greek amphitheaters also made use of similar techniques.

Beginning earlier, the acoustical traditions of the synagogue favored speech over music. In fact, music largely consisted of the cantor, a soloist who intoned various prayers and scriptures. The worship spaces tended to be on the small side, thus aiding the spoken word. Centuries later, the Protestant denominations adopted a similar position.

Today, new social styles have entered the picture, and shorter reverberation times are the norm. There is still some conflict between the requirements of speech and those of music, but reinforcement systems have for the most part made it possible for both to coexist. The biggest problem today is the high cost of building a large worship space that can do justice to the musical traditions of past centuries, and this represents a challenge for electronic acoustics, as discussed in section 11.3. In fact, the emerging art of electronic acoustics may find far greater opportunities for application in houses of worship than in concert halls.

An example of traditional church architecture in modern form is shown in figure 11-24. Saint Basil's Roman Catholic Church in Los Angeles is constructed of massive concrete piers that frame stained glass windows. A wooden ceiling rests on the piers. Internal volume is approximately 12,700 m³, and the midband (500-Hz) reverberation time in excess of 5 sec poses great problems of speech intelligibility, with or without sound reinforcement.

This church was built in 1969, and it is doubtful whether such a structure would be built during the final decade of the twentieth century, primarily because of changing views in the Roman Catholic Church. The thrust of the Vatican II reforms has emphasized the importance of speech communication and congregational involvement in the liturgy to such an extent that contemporary Roman Catholic design has much in common with Protestant and Evangelical worship design.

Such a design is shown in figure 11-25. Saint Therese Roman Catholic Church of Portland, Oregon, is built in a quasi-fan shape so that more congregants are closer to the altar. Reverberation times are generally in the 2-sec range, making normal speech reinforcement an easy task. Such a design could be typical of any Christian or Jewish house of worship built in the last two decades of this century.

In general, there is a commonality of goals in the design of houses of worship today. These are:

1. Innovative architecture, aimed at providing good sight lines from all seating areas.

2. A feeling of height and vertical dimension, often through the use of sloped ceilings and other departures from rectilinear detail.

3. Reverberation times in the range of 1.5 to 2.5 sec, depending on the musical traditions of the denomination.

4. Skillful integration of adjacent spaces for congregation overflow.

5. Acoustical design that stresses liveness for the sake of congregational singing.

6. Reliance on complex electroacoustical systems for speech and music reinforcement when needed.

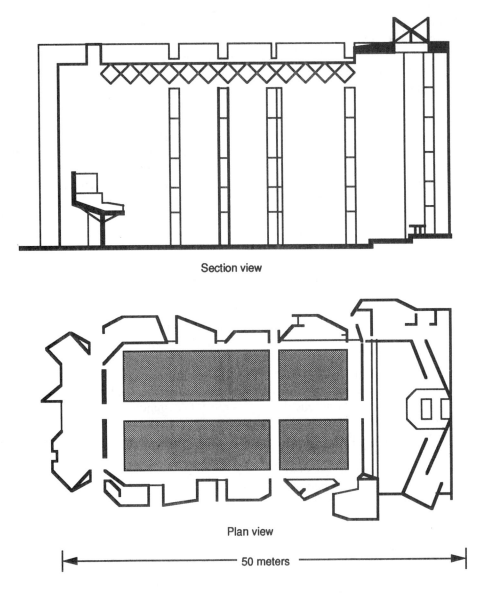

Section view

Plan view

50 meters

Figure 11-24. Views of St. Basil's Roman Catholic Church, Los Angeles, CA.

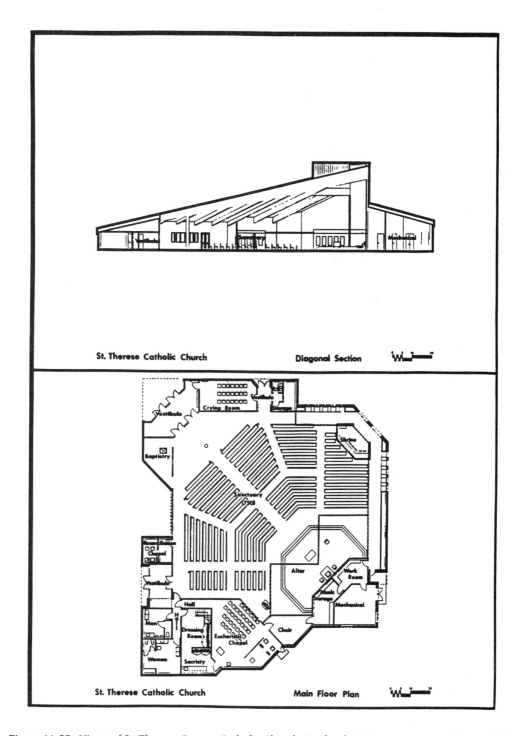

Figure 11-25. Views of St. Therese Roman Catholic Church, Portland OR. Data courtesy of ASA and Daly Engineering Company, Beaverton, OR.

11.8   ACOUSTICAL ISOLATION AND NOISE CONTROL

All the spaces studied in this chapter benefit from proper acoustical isolation from outside sources and control of potential local noise sources. In fact, most of the work done by acoustical consultants is in these areas, as opposed to the purely creative pursuit of providing "good acoustics" for the client.

We have already discussed at length the acoustician's job of controlling the absorption and reflection of sound within a space; now we will discuss the problems he faces with regard to interfering noises.

First, there is the matter of priority. In a concert hall, where patrons have paid dearly for tickets, there is no excuse for intrusion from overflights or fire sirens. It is only at great expense that the house is properly isolated from such sounds. Carnegie Hall in New York City has suffered for years from slight intrusion from the BMT subway system, but it is apparent only during quiet musical passages. The cost of additional isolation would be prohibitive. By comparison, in a house of worship, where there may be an occasional crying child, the added intrusion of an overflight or siren may not be a problem.

The acoustician first determines the desired level of noise in the space. This is usually determined by what are called noise criterion (NC) curves, as shown in figure 11-26. These curves follow very closely the equal loudness contours shown in figure 2-2, and they are measured on octave bands. The following list shows the NC ranges for various applications.

Application	NC Rating
Quiet homes	35–45
Houses of worship	30–40
Offices	35–45
Motion picture theaters	25–35
Concert halls	10–25
Recording studios	5–20

Once a target NC rating has been set, the acoustician's job falls into two areas. First, he must ensure that there are no local noise sources that would defeat the NC goal. The concern here is mainly with air handling and other physical necessities. Elevators and plumbing can be sources of noise and must be identified and corrected at the outset. Electrical facilities such as lighting control systems can generate disturbing levels of hum. Potential impact noise sources, which are structure borne, must also be identified at this stage and eliminated.

Air handling concerns include not only turbulence at the exit registers, but also the possibility of coupling of sound from adjacent spaces that are fed by the same system. There is adequate design history and data in the air conditioning industry to enable the right choices to be made at the outset. In general, ductwork of large cross-sectional area and multiple openings into the performance space will minimize noise.

The second step is to identify the long-term nature of noises outside the building. This is done by monitoring those noise levels for some period of time and making sure that all seasonal factors have been taken into account. The acoustician may determine that, for some large percentage of the time, a given noise spectrum outside the building will not be exceeded.

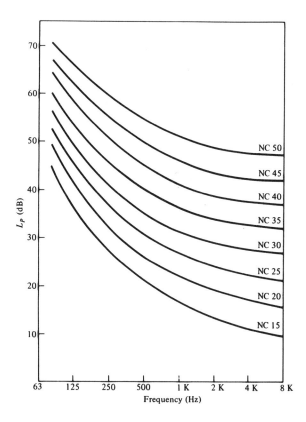

Figure 11-26. Noise criterion (NC) curves.

He then will subtract each octave band noise level from the corresponding octave band for the target NC curve and tabulate the differences between them, calculating the desired airborne noise transmission loss (TL) from the outside. The TL data are fed back to the building committee for assessment of their impact on construction costs.

The acoustician further identifies the particular nature of construction that will enable the architectural engineer to achieve the desired sound isolation. For many kinds of wall structure, these data have already been measured and exist in handbook form.

Various wall and floor structures are rated by their sound transmission class (STC). The family of STC curves is shown in figure 11-27. Note that the rating number for a given curve is based on its sound attenuation in the frequency range of 500 Hz. Above that frequency there is slightly more attenuation of sound, while below that frequency the attenuation falls off rather quickly.

As an example of how the acoustician uses the design charts, let us assume that the long-term noise levels outside a given building are as follows:

Frequency, Hz	125	250	500	1000	2000	4000
Measured noise	55	50	52	48	50	52
NC 25	45	38	31	27	24	22
Difference (TL)	10	12	21	21	26	30

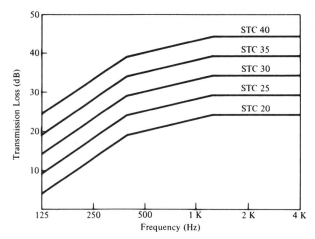

Figure 11-27. Sound transmission class (STC) curves.

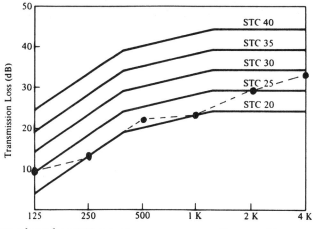

Figure 11-28. TL data plotted on STC curves.

The TL data are determined by subtracting the measured noise levels from the desired NC requirements. These data are then plotted on the the STC curves, as shown in figure 11-28, and the acoustician notes the lowest STC curve that the TL data do not exceed by more than 2 dB (averaged over the entire range). In this case, it is clear that STC 20 to 25 will satisfy the desired sound isolation requirements.

It is the acoustician's responsibility to determine all possible sources of noise problems. As one set of problems is solved, another unforeseen set may become evident. In particular, after much attention has been paid to wall structure, other flanking paths, such as floor coupling, may become significant.

In most concert halls, there is usually enough in the way of isolation of the hall, through buffer zones, hallways, and the like, that outside noises are not likely to be a problem. Air handling is usually the most expensive item on the list of problems to be solved. In the case of rehearsal spaces in large complexes, airborne noise may be the most significant factor, because of the many adjacent activities.

REFERENCES

Ando, Y. 1985. *Concert Hall Acoustics*. New York: Springer-Verlag.

Barron, M. 1971. "The Subjective Effects of First Reflections in Concert Halls—The Need for Lateral Reflections." J. *Sound and Vibration* 15:475–494.

Benade, A. 1976. *Fundamentals of Musical Acoustics*. New York: Oxford University Press.

Beranek, L. 1962. *Music, Acoustics & Architecture*. New York: Wiley.

Borish, J. 1984. "Extension of the Image Model to Arbitrary Polyhedra." J. *Acoustical Society of America* 75:1827–1836.

Brook, R. 1987. "Rooms for Speech, Music, and Cinema." Chap. 7 in *Handbook for Sound Engineers*. Indianapolis, IN: H. Sams.

Cremer, L., et al. 1978. *Principles and Applications of Room Acoustics*. Vol. 1. New York: Applied Science Publishers.

Doelle, L. 1972. *Environmental Acoustics*. New York: McGraw-Hill.

Fitzroy, D. 1959. "Reverberation Formula Which Seems to be More Accurate with Nonuniform Distribution of Absorption," *Acoustical Society of America*, 31:893–897.

Forsyth, M. 1985. *Buildings for Music*. Cambridge, MA: MIT Press.

French, N., and J. Steinberg. 1947. "Factors Governing the Intelligibility of Speech Sounds." J. *Acoustical Society of America* 19:90–119.

Houtgast, T., and H. Steeneken. 1972. "Envelope Spectrum and Intelligibility of Speech in Enclosures." Paper presented at IEE-AFCRL Speech Conference.

Knudsen, V., and C. Harris. 1950. *Acoustical Designing in Architecture*. New York: Wiley.

Kryter, K. 1962. "Methods for the Calculation and Use of Articulation Index." J. *Acoustical Society of America* 34:1689.

Kuttruff, H. 1979. *Room Acoustics*. London: Applied Science Publishers.

Parkin, P. 1975. "Assisted Resonance." In *Auditorium Acoustics*, ed. R. Mackenzie. London: Applied Science Publishers.

Peutz, V. 1971. "Articulation Loss of Consonants as a Criterion for Speech Transmission in a Room." J. *Audio Engineering Society* 19(11).

Philips Product Bulletin: Multichannel Reverberation System (published by Audio-Video Systems group).

Pierce, J. 1983. *Musical Sound*. New York: W. H. Freeman.

Reichardt, W., A. Alim, and W. Schmidt. 1974. *Applied Acoustics*, vol. 7.

Schroeder, M. 1984. "Progress in Architectural Acoustics and Artificial Reverberation." J. *Audio Engineering Society* 32(4).

Schultz, T. 1965. "Acoustics of the Concert Hall." IEEE *Spectrum*.

Thiele, R. 1953. *Acustica* 3:291.

Veneklasen, P. 1975. "Design Considerations from the Viewpoint of the Professional Consultant." In *Auditorium Acoustics*. London: Applied Science Publishers.

ADDITIONAL RESOURCES

Acoustics of Worship Spaces. Acoustical Society of America, New York: 1985.

Anechoic Orchestral Music Recording, Denon Compact disc PG-6006.

Halls for Music Performance. Acoustical Society of America, New York: 1982.

Theatres for Drama Performance. Acoustical Society of America, New York: 1986.

12

Principles of Speech and
Music Reinforcement

The chief goal of speech reinforcement is to increase intelligibility in environments where unaided speech cannot easily be understood. The primary requirements for intelligibility as addressed by the reinforcement system engineer are:

1. Maintaining an adequate speech signal level

2. Maintaining an adequate speech-to-noise ratio

3. Minimizing the effects of reverberation and discrete sound reflections

The first of these requirements is easily met through proper hardware specification and through proper talker-microphone positioning. The second requirement is a bit more complex in that noise may be considered as both random and signal-related. Random noise in the audience area may be due to air handling, outside traffic, or sounds produced by the audience itself. The effect of reverberation in the acoustical space is to create a kind of noise behind the signal which, under certain conditions, will impede intelligibility. Raising the speech level in the space will provide some immunity to random noise; however, the effects of reverberation will follow the speech signal itself, remaining at a fixed level below it. The goal of the system engineer is to provide uniform coverage over the audience area, while at the same time maintaining or increasing the direct-to-reverberant ratio through the use of loudspeakers that aim acoustical power precisely where it is wanted.

12.1 BASIC APPROACHES

In most large spaces, such as auditoriums or houses of worship, the normal approach is to place a central loudspeaker array overhead toward the front. Speech originating from such a position will sound quite natural, since it is perceived as coming from the same azimuthal direction as the talker. However, in highly reverberant spaces, a central array may not provide an adequate direct-to-reverberant ratio; in such cases, multiple distributed loudspeakers located closer to the audience may be required to attain the desired ratio. If the designer has a

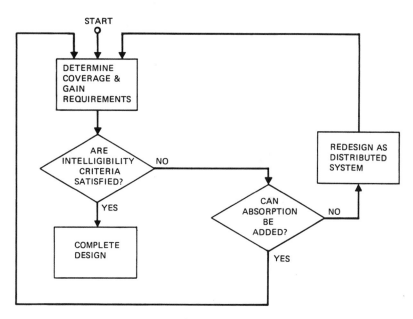

Figure 12-1. The basic iterative process in the design of a sound reinforcement system.

hand in the acoustical treatment of the space, then it may be possible to add more acoustical absorption, which will lessen the reverberation time and thus reduce the direct-to-reverberant ratio. Another set of calculations can then be made to determine whether the central array will work. The iterative design process is shown in the operational flow diagram of figure 12-1.

The job is not an easy one, and the designer often has to balance performance requirements within fixed budget constraints. A properly designed distributed system will generally cost more than a central array, and some degree of value engineering may be required to sort out all the available options.

12.1.1   Typical Block and Signal Flow Diagrams

Figure 12-2a shows the block and signal flow diagrams for a central array. The triangle indicates a power amplifier, and the dividing network splits the signal into separate high- and low-frequency components.

In figure 12-2b we show a similar system, this time with biamplification. Biamplification provides for separate powering of the high- and low-frequency loudspeakers, resulting in less loudspeaker distortion at high power input.

In figure 12-2c we show the flow diagram for a distributed system. Note that sets of loudspeakers farther away from the talker have been progressively delayed so that sound arriving at the listener from the loudspeakers is essentially "in step" with that arriving directly from the talker. When the delays have been properly adjusted, the effect to a distant listener is that the reinforced sound is heard as originating from the actual source. Recall the discussion of the precedence effect in section 2.3.2.

It is possible to combine the virtues of both a central array and a distributed system, as we shall see later.

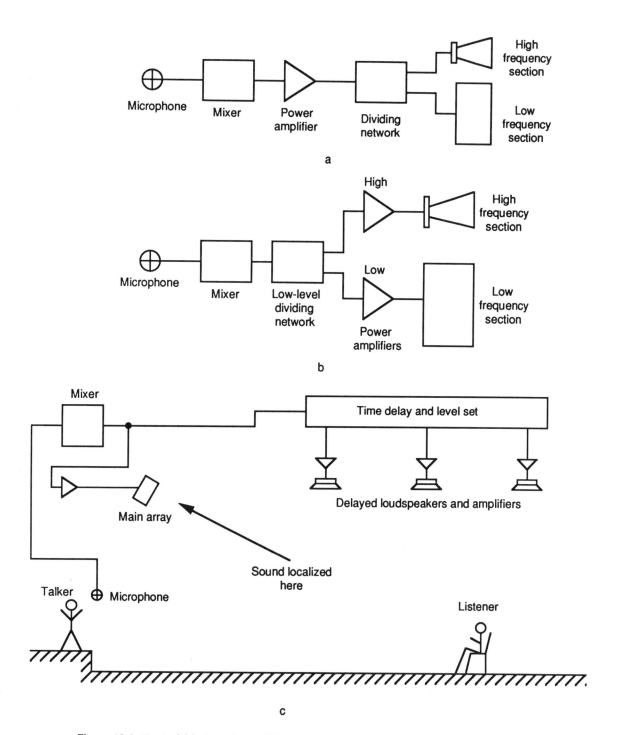

Figure 12-2. Typical block and signal flow diagrams. (*a*) Central array; (*b*) central array biamplified; (*c*) distributed system.

12.1.2   Loudspeaker Hardware Requirements

In general, the loudspeaker components used in high-level sound reinforcement are different from those that might be used in the home. They are rugged and capable of high acoustical output. The high-frequency units that are generally preferred are of the horn type, since the horn can be designed for precise pattern control. Figure 12-3 shows data on a typical uniform coverage horn, which covers the frequency range from 500 Hz upward. A line drawing of a 90° by 40° horn is shown in figure 12-3a. The −6-dB beamwidth of the horn is shown in figure 12-3b, and the frontal isobars are plotted in figure 12-3c.

Many manufacturers offer similar devices, and common patterns, in addition to the 90° by 40°, are 60° by 40° and 40° by 20°. The directivity indices of these horns are typically in the following range.

Nominal pattern	Directivity index, dB
90° by 40°	10–12
60° by 40°	12–14
40° by 20°	15–17

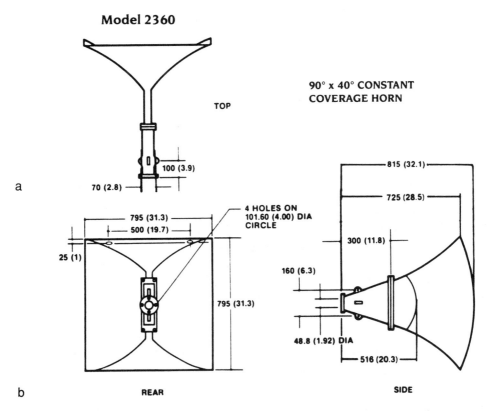

Figure 12-3. A typical uniform coverage horn. (a) Line drawing; (b) plot of −6-dB beamwidth; (c) frontal isobars. Data courtesy of JBL Inc.

Low-frequency requirements are normally met with arrays of cone loudspeakers mounted in ported boxes. Pattern control at low frequencies is difficult to maintain; fortunately, the requirements for it are relaxed, as speech articulation depends primarily on system performance in the range from 500 Hz to 4 kHz.

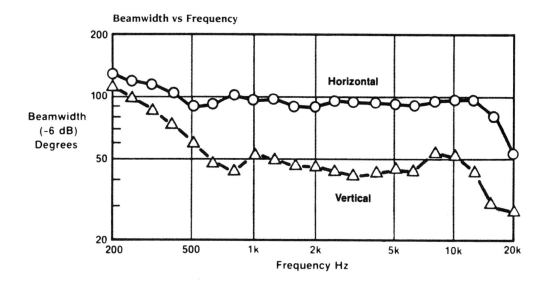

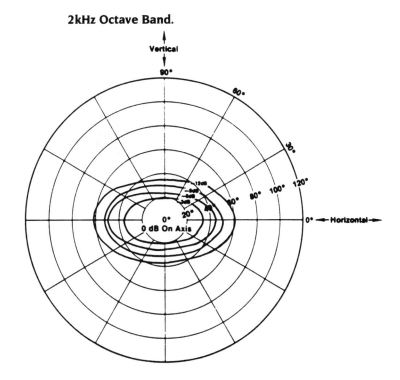

12.1.3   Requirements for Proper Coverage

The designer must first ensure that the entire seating area is properly covered. This is done by carefully choosing high-frequency horns and splaying them so that their patterns merge properly over the entire area. Drive levels are then adjusted for uniform coverage over the seating area. Computer programs that simplify the aiming process are available; however, the choice of componentry is still in the hands of the designer.

When more than one device is used to cover an area, there will inevitably be wavelength-related interferences between them. These are referred to as *lobes*, because of their appearance in polar diagrams. Lobing can be minimized if the overlap zones between adjacent radiators are chosen at angles where the response is changing rapidly. In other words, the coverage angles of devices should overlap only to the degree necessary to afford smooth coverage.

Ideally, the designer wishes to direct the coverage only at the audience, with as little sound as possible striking the walls or other reflective surfaces in the room. Figure 12-4 illustrates what happens to the reverberant level in a room as a function of the absorption coefficient of the surface upon which the sound is first incident. If sound is aimed at a reflective surface, little of it is absorbed, and the resulting reverberant level will be fairly high. If the surface of first incidence has a high absorption coefficient, then considerable sound will be absorbed initially, and there will be little left to contribute to the reverberant level in the room. Considerations such as these all but rule out the use of omnidirectional or other low-directivity devices in central arrays. However, there are many applications in distributed systems where low directivity hardware is quite adequate.

12.2   SYSTEM STABILITY AND GAIN CALCULATIONS

Perhaps the most common problem in sound system operation is electroacoustical feedback. Feedback occurs when the gain of the system is too high, so that sound from the loudspeaker enters the microphone and recirculates through the entire system. The maximum gain the system can attain is determined by the various distances between talker, microphone, loudspeaker, and listener. The acoustical nature of the room is important here as well.

Indoor systems are complicated, but a simple outdoor system can be analyzed fairly easily. Referring to figure 12-5, note the four paths indicated between talker, microphone, listener, and loudspeaker. Acoustical gain is defined as the difference in level that the listener perceives with the system turned on, compared with the level from the talker when the system is off. The following equation gives the potential acoustical gain that the system can attain:

$$\text{Potential gain} = 20 \log D_1 - 20 \log D_2 + 20 \log D_o - 20 \log D_s - 6 \qquad (12.1)$$

The term in the equation that is most easily varied is the distance between talker and microphone, D_s. Since gain is inversely proportional to this distance, halving the distance has the effect of increasing the system's potential gain by 6 dB. The other terms in the equation are more or less fixed and cannot be easily changed.

Other factors to be considered include the number of open microphones in the complete sound reinforcement system. Each microphone provides its own path through the system, and thus increases the chances of feedback. In proper system operation, unused microphones are muted, either automatically or manually by an operator. Directional microphones and loud-

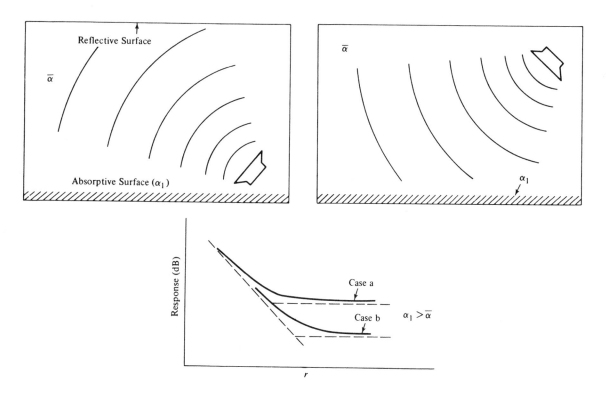

Figure 12-4. Reverberant level as a function of the absorption coefficient of the surface the sound is first incident upon.

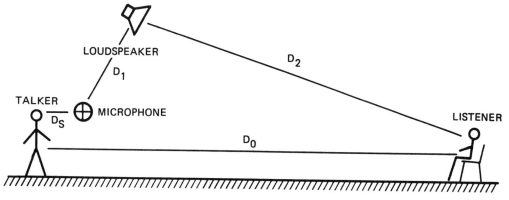

Figure 12-5. Basic elements of an outdoor sound reinforcement system.

speakers can increase the potential gain of the system by a maximum of about 5 or 6 dB. In the form shown here, equation 12.1 assumes that both microphone and loudspeaker are omni-directional. The final term in the equation, −6 dB, represents a safety factor. If a system is operated at its upper limit of stability, there may be some degree of "ringing" as the feedback mechanism tries to go to work. Reducing the gain by 6 dB will cure this.

As an example of system potential gain calculation, let us assume the following conditions:

$$D_s = 1 \text{ m}$$

$$D_o = 7 \text{ m}$$

$$D_1 = 4 \text{ m}$$

$$D_2 = 6 \text{ m}$$

Then, by equation 12.1,

$$\text{Potential gain} = 17 - 0 + 12 - 15.6 - 6$$
$$= 7.4 \text{ dB}$$

12.3   INTELLIGIBILITY CALCULATIONS

The general requirements for speech intelligibility were discussed in section 11.4.2, and the designer of a sound reinforcement system must be aware of these requirements and ensure that the system will meet them. To the extent that the designer can estimate reverberation time, reverberant levels in the space, and the effects of random noise, he will be in a position to make an informed estimate of just how well the system will work.

In the case of auditoriums, where outside noises are minimal, the system designer will probably make use of the Peutz criteria for system intelligibility. For a given listening position in the hall, the designer will calculate the direct level at the listener as produced by the loudspeaker. Then, the reverberant level will be calculated, based on the total acoustical power output of the system. Finally, the reverberation time in the hall will be calculated, or estimated.

With this information, the designer can refer to the graph shown in figure 11-22, and estimated intelligibility can be related to the direct-to-reverberant ratio and the reverberation time.

If the designer was working on a system to be installed in a noisy transportation terminal where reverberation time was quite short, the articulation index (AI) would yield a more accurate estimate of system intelligibility.

The greatest problems the designer has in estimating system performance before either the room or the system has been built have to do largely with the unexpected, such as strong discrete echoes, unaccounted-for sources of noise, and the like. At best, this is a risky process, and most designers proceed very cautiously, checking their work at every turn.

12.4   SYSTEMS FOR HOUSES OF WORSHIP

Figure 12-6 shows six approaches to speech reinforcement system design suitable for reverberant houses of worship. The general architecture here is typical of church design through the first sixty or so years of the twentieth century, when structures were long, narrow, and relatively high. The specific approach to system design depends on the amount of reverberation in the space.

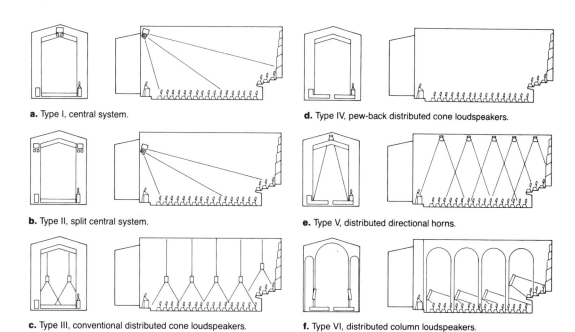

a. Type I, central system.

b. Type II, split central system.

c. Type III, conventional distributed cone loudspeakers.

d. Type IV, pew-back distributed cone loudspeakers.

e. Type V, distributed directional horns.

f. Type VI, distributed column loudspeakers.

Summary of typical sound reinforcement systems for worship spaces.

	TYPE OF SYSTEM	APPLICATION	DESIGN CONSIDERATIONS	DELAY UNIT
I.	Central directional cluster of horns (sometimes column loudspeakers for "easy" systems)	Where architecture permits	Large radiating area required for directional control, line-of-sight to all listeners, lack of distant sound-reflecting surfaces to produce echoes. Higher reverberation time requires more directional control and larger radiating area	Not required
II.	Split directional cluster of horns (columns for "easy" systems)	Where most speech originates from left and right (for example, pulpit and lectern)	Same as above. In addition, the lectern signal should usually be amplified through its loudspeaker only; and the pulpit through its loudspeaker only	Not required
III.	Conventional distributed system–cones directed vertically	Low-ceilinged spaces; under-balcony areas; where direct sound is at a minimum	Loudspeaker sufficiently low 4.5 meters (15 ft) maximum in reverberant spaces. Consider chandeliers. Close enough on-center spacing for even coverage and loudspeakers with wide treble coverage	Essential when supplementing a main directional system; otherwise essential for directional realism and highest intelligibility
IV.	Pew-back distributed small cones	Where other systems are not applicable (expensive)	Large number of loudspeakers, one per three listeners; small loudspeakers high on backs, never under pews	Essential for directional realism and highest intelligibility, especially where live sound is strong
V.	Distributed directional horns	Hard cases with no sound-absorption other than people and where sound should be confined to occupied areas	Large single directional horns directed vertically, each covering relatively small precisely determined areas. Loudspeakers should be no higher than 13 meters (45 ft)	As above
VI.	Distributed column loudspeakers	Long narrow spaces where columns provide logical mounting locations	Distance between left and right columns no greater than 13 meters (45 ft), columns tilted to provide defined coverage, best results with custom-designed column loudspeakers	Always required

Figure 12-6. Approaches to sound reinforcement in reverberant houses of worship. Data courtesy of AES and D. L. Klepper 1970.

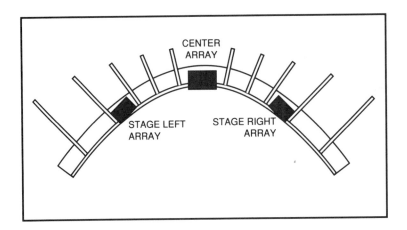

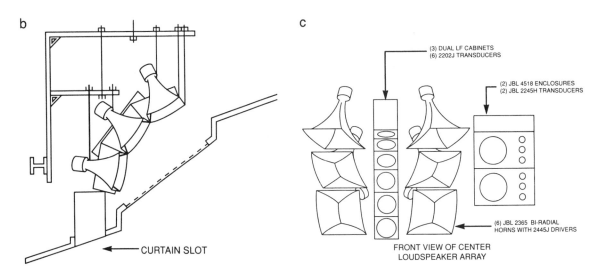

Figure 12-7. Loudspeaker array for a stereophonic system in a house of worship. (*a*) Plan view; (*b*) side view of center array; (*c*) front view of center array. Data courtesy of Klepper 1970 and JBL Inc.

Recent architecture has emphasized fan-shaped structures with lower reverberation times. Changing worship styles have emphasized stereophonic music reinforcement as well. Figure 12-7 shows a view of a system designed for a fan-shaped house of worship.

12.5   SYSTEMS FOR AUDITORIUMS

At one time, auditoriums were used principally for musical presentations, and sound reinforcement was not a necessity. Today, spaces whose primary function is music presentation may be widely used for other purposes, and sound reinforcement systems are a necessity. While many traveling musical groups bring in their own portable sound reinforcement gear, it has become customary to install a permanent system in the hall. A typical approach is shown in figure 12-8. The main array is located above the proscenium. Pairs of high-frequency horns are used for near, mid, and far coverage. Low-frequency coverage is provided by a vertical arc array consisting of six transducers. The far sides at the front are slightly outside the main coverage angles of the near horns, and small loudspeakers located at the proscenium sides may be used for added coverage. Under-balcony coverage may require delayed overhead loudspeakers.

Figure 12-9 shows details of a horizontal line array extending the entire width of a rectangular hall. Such an array provides excellent coverage from front to back in the hall. Attenuation of direct level with distance from a line array is 3 dB per doubling of distance (see Section 1.7.2), and this ensures excellent intelligibility by maintaining an adequate direct-to-reverberant ratio throughout the house.

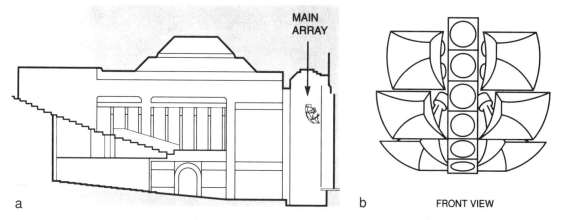

Figure 12-8. Central array in an auditorium. (*a*) Elevation view of auditorium; (*b*) front view of array. Data from Klepper 1970 and JBL Inc.

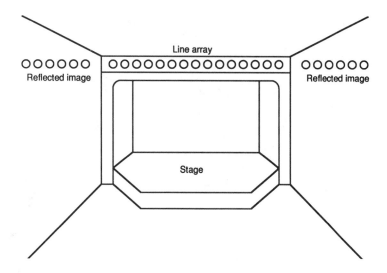

Figure 12-9. Horizontal line array in a concert hall.

Figure 12-10. Main and stage monitor systems for large-scale music reinforcement. Photo courtesy of JBL Inc.

12.6   SYSTEMS FOR HIGH-LEVEL MUSIC REINFORCEMENT

The special needs of high-level music reinforcement include:

1. Smooth system power bandwidth capability over the required frequency range (normally from 30 Hz to 10 kHz).

2. Sufficient overall acoustical output capability within given distortion limits (often in excess of 110 dB L_p).

3. Sufficient loudspeaker elements to direct the sound toward all patrons, as required.

An additional requirement is *stage monitoring,* in which sound is directed back toward the musicians so that they can maintain proper ensemble.

The requirements for speech intelligibility are often waived in favor of purely musical requirements. In an effort to give all patrons sufficient coverage, many loudspeakers may be aimed in many directions. This may result in direct-to-reverberant ratios that do not favor high speech intelligibility. Additionally, in large venues such as sports arenas and the like, the result may be multiple signals reaching many listeners, with substantial acoustical delay between them. Such effects are not always detrimental to music; however, very large delays are often adjusted, to the degree possible, by the use of electronic signal delay devices.

Figure 12-10 shows a photograph of the main array and monitor systems typical of a large rock music performance.

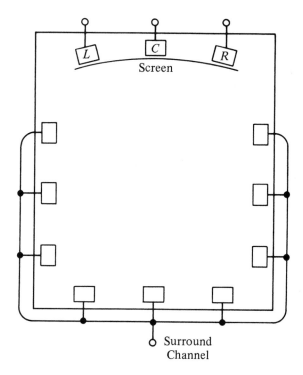

Figure 12-11. Basic layout for a multichannel motion picture system.

12.7   SYSTEMS FOR MOTION PICTURE THEATERS

Since the early 1980s, special attention has been paid to the electroacoustical requirements of motion picture exhibition. Such systems are normally multichannel, with left-center-right presented through the screen and a surround channel in the house wrapped around the patrons and presented over multiple loudspeakers. Details are shown in figure 12-11. The signals used to drive the multichannel loudspeaker array may be discrete, via magnetic tracks on the film, or they may be via a matrixed two-channel system recorded optically.

The screen channels should each be capable of producing a peak output of 95 to 100 dB L_p toward the middle of the house, and the surround channel, taken as an ensemble, should be capable of a similar output. Very low frequencies (in the range from 25 to 40 Hz) may be augmented by a set of *subwoofers* located behind the screen, normally driven by a summed monophonic signal from all channels. The surround loudspeakers may be designed only for performance down to about 50 Hz, thus minimizing their size requirements.

High-frequency losses through the perforated screen are substantial above about 5 kHz, requiring high-frequency boosting if standardized frequency response is to be maintained in the house.

REFERENCES

Ballou, G. 1987. *Handbook for Sound Engineers*. Indianapolis: H. Sams.

Boner, C., and R. Boner. 1969. "The Gain of a Sound System." J. *Audio Engineering Society* 17(2).

Davis, D., and C. Davis. 1987. *Sound System Engineering*. Indianapolis: H. Sams.

Eargle, J. 1989. *Handbook of Sound System Design*. Commack, NY: Elar Publishing Co.

———, J. Bonner, and D. Ross. 1985. "The Academy's New State of the Art Loudspeaker System." J. *Society of Motion Picture and Television Engineers* 94(6).

Klepper, D. 1970. "Sound Systems in Reverberant Rooms for Worship." J. *Audio Engineering Society* 18(4).

Various. 1978. "Sound Reinforcement." Compiled from the pages of the *Journal of the Audio Engineering Society*, New York.

13

Principles of Sound Recording

As a commercial enterprise, sound recording dates from the late 1870s. Edison's crude demonstration of the reproduction of speech sounds indented on tinfoil led to an industry that thrived despite severe limitations on acoustical bandwidth, noise, and distortion. For the first fifty years the art and science of recording and playback remained entirely acoustomechanical in nature.

The introduction of electrical recording by Western Electric (Maxfield and Harrison 1926) during the late 1920s provided more artistic freedom for musicians, engineers, and record producers, and the groundwork was laid for the world of in-studio sound creation that we see today in so much popular and rock music. For classical music, the advantages of electrical recording were that singers and players did not have to produce such high acoustical levels in order to modulate a phonograph record groove directly. They could perform at their normal levels, and with subtlety, knowing that the microphone could capture it all.

During the early 1930s, experiments in multichannel recording were undertaken in England (Blumlein 1958) and in the United States (Steinberg and Snow 1934). These led to the formulation of techniques for recording auditory perspectives. The term *stereophonic*, or simply *stereo*, is used today to describe this. Stereo reached the consumer's living room during the 1950s in the form of reel-to-reel recorded tapes and stereo long-playing (LP) records.

Today, we enjoy stereo primarily through the compact disc (CD), a digital medium, and the Philips cassette, a convenient analog tape medium, noting with some nostalgia that the stereo LP seems to be on the way out.

In this chapter we will discuss primarily the tools and techniques of sound recording, reserving for a later chapter the description of home implementation of audio technology.

13.1 BASIC SYSTEM CONCEPTS

While most readers of this book will already have some knowledge of how a recording chain is put together, there will be some who need a primer here. Figure 13-1a shows the basic acoustical recording and playback system as invented by Edison. Sound is gathered by the

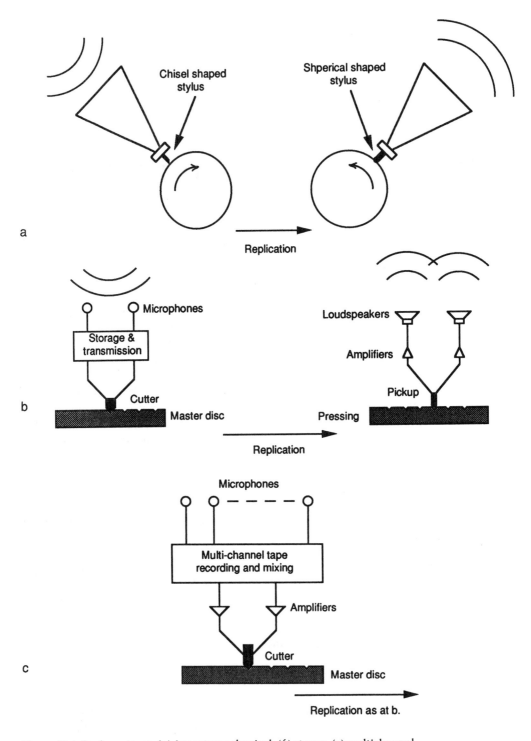

Figure 13-1. Basic systems: (*a*) Acoustomechanical; (*b*) stereo; (*c*) multichannel.

recording horn, and low pressure at the mouth of the horn is transformed into high pressure at the throat, where it can efficiently move a diaphragm. A chisel-shaped stylus attached to the diaphragm is placed in a rotating plastic medium, and the vibrations of the stylus cut a modulated groove into the cylindrical surface. The cylinder is replicated by plating and subsequent molding, and the recording is ready for playback. The playback process is basically the inverse of the recording process.

The basic two-channel stereo chain shown in figure 13-1*b* benefits from artistic choices of gain adjustment, equalization, and editing, all of which are in the hands of the recording engineer and record producer. Recording and playback are by magnetic tape, and subsequent electrical transfers allow replication in either disc or tape format for home use.

The multichannel chain shown in figure 13-1*c* provides for storage of all the studio ingredients for later remix. At that time, various types of signal processing can be used to modify, or otherwise enhance, the signal, and a master stereo recording is created. Further replication is the same as with the system shown in figure 13-1*b*.

13.2   MICROPHONES

The microphone is the first element in the complex electroacoustical chain linking the studio with the loudspeaker in the listening room. Our main concerns here will be pickup patterns, basic methods of operation, departures from ideal performance, and pertinent microphone specifications.

13.2.1   Microphone Pickup Patterns

There are three primary pickup patterns: omnidirectional (sound pickup is essentially uniform in all directions), unidirectional (sound pickup is primarily on one side of the microphone), and bidirectional (sound pickup is predominantly from the front and back of the microphone). In the bidirectional pattern, the front and back lobes are of opposite polarity; thus plus and minus signs are indicated in the diagram. Figure 13-2*a*, *b*, and *c* shows these patterns in two-dimensional form. The reader should realize, however, that the patterns exist in three dimensions.

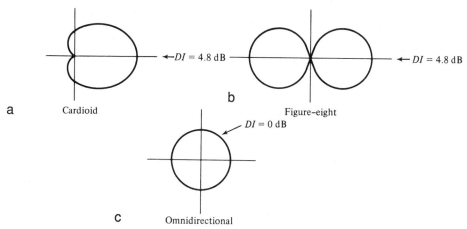

Figure 13-2. Basic microphone patterns: (*a*) Unidirectional; (*b*) bidirectional; (*c*) omnidirectional.

The unidirectional pattern is often referred to as the *cardioid* pattern, the name deriving from its heart shape. The bidirectional pattern is often referred to as the *cosine* pattern, since it resembles a plot in polar coordinates of the cosine function in mathematics. The term *figure-8* is also used to describe this pattern.

The cardioid pattern can be derived from a combination of omnidirectional and bidirectional patterns, producing a family of what are called *first-order patterns*. These are all useful to the recording engineer, and some microphone models containing dual elements are capable of being switched from one of these patterns to another.

The data shown in figure 13-3 give the principal characteristics of some of the first-order patterns. In this figure, the primary pickup direction is assumed to be toward the top of the figure, and that direction is referred to as the *primary axis*. (The term *on-axis* is often used by recording engineers to indicate microphone performance, or placement of musical sources, along the major axis.) Most of the characteristics presented in figure 13-3 are self-explanatory; however, two of them are not obvious.

The *random efficiency* is a measure of the on-axis response relative to the sum of the response in all directions. For the omnidirectional pattern, which is taken as a reference, this is unity. For the bidirectional and unidirectional patterns, reverberant pickup will be only one-third as

CHARACTERISTIC	OMNIDIRECTIONAL	BIDIRECTIONAL	CARDIOID	HYPERCARDIOID	SUPER-CARDIOID
POLAR RESPONSE PATTERN	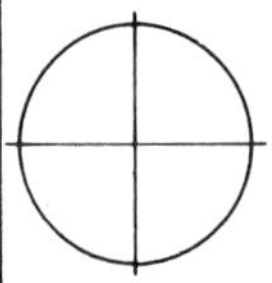	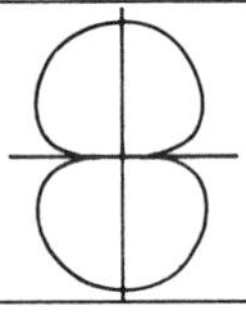	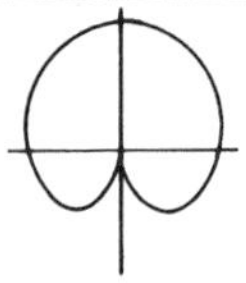	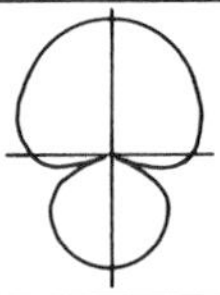	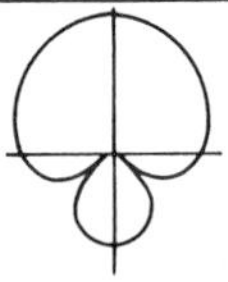
POLAR EQUATION F (θ) ∝	1	$\cos \theta$	$1/2(1+\cos \theta)$	$1/4(1+3\cos \theta)$	$.37+.63\cos \theta$
PICKUP ARC 3 dB DOWN (θ3)	360°	90°	131°	105°	115°
PICKUP ARC 6 dB DOWN (θ6)	360°	120°	180°	141°	156°
RELATIVE OUTPUT AT 90° (dB)	0	$-\infty$	−6	−12	−8.6
RELATIVE OUTPUT AT 180° (dB)	0	0	$-\infty$	−6	−11.7
ANGLE AT WHICH OUTPUT=0 (θ₀)	—	90°	180°	110°	126°
RANDOM ENERGY EFFICIENCY (RE)	1 0 dB	.333 −4.8 dB	.333 −4.8 dB	.250 ① −6.0 dB	.268 ② −5.7 dB
DISTANCE FACTOR (DSF)	1	1.7	1.7	2	1.9

① MINIMUM RANDOM ENERGY EFFICIENCY FOR A FIRST ORDER CARDIOID
② MAXIMUM FRONT TO TOTAL RANDOM ENERGY EFFICIENCY FOR A FIRST ORDER CARDIOID

Figure 13-3. First-order microphone characteristics. Courtesy Shure Brothers.

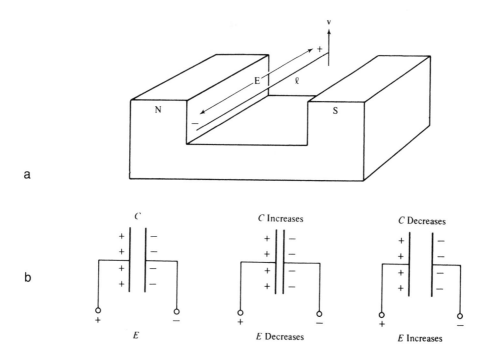

Figure 13-4. Operating principles of microphones: (a) Magnetic induction; (b) capacitive.

great as for the omnidirectional pattern for equal on-axis response. This implies that these two patterns will have more on-axis "reach" than will the omnidirectional pattern. In practical terms, this means that microphones with those patterns can be used in reverberant spaces, as they provide more isolation from the reverberant field than can the omnidirectional pattern.

Distance factor, which is equivalent to the term directivity factor, or Q, of a loudspeaker, is another way of looking at the same characteristic. For example, if a hypercardioid microphone and an omnidirectional microphone are both used to pick up a sound source located in a reverberant field, the hypercardioid microphone may be operated twice as far away from a given source and still pick up no more reverberation than the omnidirectional microphone. It is essential that the primary axis of the hypercardioid microphone be aimed at the source for this to be the case.

13.2.2   Basic Operating Principles

Modern professional microphones operate on one of two physical principles: magnetic induction and variable capacitance. These principles are shown in figure 13-4a and b, respectively. Microphones using the magnetic induction principle employ a fixed magnet and a movable diaphragm, to which is attached a small, lightweight coil that is placed in the magnetic field. When sound impinges on the diaphragm, it moves slightly, and a voltage is generated in the coil. The voltage E is given by the following equation:

$$E = Blv \tag{13.1}$$

In this equation, B is the magnetic flux density (tesla), l is the length of the coil of wire (meters), and v is the effective (rms) velocity of the diaphragm (meters per second).

Through careful acoustical damping and control of cavity resonances, extended response can be attained with a moving coil microphone. This design is often referred to as a *dynamic microphone*.

Microphones operating on the variable capacitance principle are called *capacitor microphones*. The term "condenser" microphone is a holdover from the past. In this design, an electric charge is held between the fixed backplate and the movable diaphragm. Such a charge may be provided by an external bias voltage supply or, more simply, through the use of a prepolarized electret backplate. For a given charge Q between the electrodes, the resulting voltage between the plates will be inversely proportional to the capacitance C between them. Thus, when the separation between electrodes varies as a result of sound pressure variations, the output voltage will vary according to the following equation:

$$E(\text{output signal}) = \frac{Q}{\Delta C} \qquad (13.2)$$

where ΔC is the change in capacitance caused by diaphragm motion.

The signal from the electrodes is normally amplified directly at the microphone itself because of susceptibility to signal losses and noise pickup.

Capacitor microphones are inherently simple in construction and are generally the choice of recording engineers when smooth and extended response is desired.

The *ribbon microphone* is not as common as the other two types. It operates on the principle of magnetic induction and consists of a corrugated ribbon mounted in a magnetic field. Because of the very short length of conductor in the magnetic field, the output voltage is quite small and is generally raised by means of a transformer located in the microphone case.

Both dynamic and capacitor microphones take many forms, and single-diaphragm designs are available that produce both omnidirectional and cardioid response patterns.

13.2.3   Departures from Ideal Performance

Figure 13-5a shows the proximity effect of a typical bidirectional microphone when it is operated at a distance of 0.6 m (2 ft). The rise at frequencies below about 200 Hz is a simple consequence of physics, not a response aberration as such (Eargle 1982). The effect can be put to good use in popular recording, where it can add a pleasing low-frequency response rise to vocals. At greater distances, shown here as 1.2 and 2.4 m, the low-frequency rise due to proximity is minimal. At greater operating distances in classical recording, it is not a problem at all.

Figure 13-5b shows typical off-axis falloff at high frequencies for a small diaphragm omnidirectional microphone. Obviously, the microphone is not omnidirectional at high frequencies, despite its published description! Recording engineers need to know this, and the general rule is to point the axis of maximum high-frequency sensitivity in the direction of the ensemble or soloist being recorded. For test and measurement purposes, microphones with a frequency response that is smoothest when used at off-axis angles may be designed. The intent in these designs is to provide the smoothest pickup of random incidence signals.

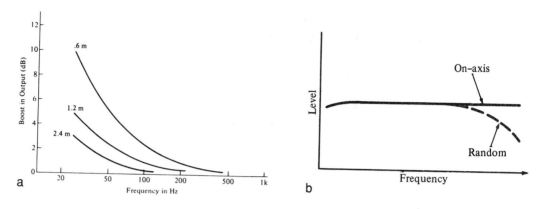

Figure 13-5. Departures from ideal microphone response: (a) Proximity effect; (b) on-axis beaming.

All microphones, whatever their directional type, will exhibit some degree of high-frequency pattern narrowing along the major axis. There may be some degree of pattern widening at low frequencies as well, depending on the actual design of the microphone. The user must be aware of this and compensate accordingly. Recording engineers speak of these anomalies as "off-axis coloration," and the term refers to the fact that the frequency response of sounds picked up off-axis will be modified by the degree of pattern deviation that the microphone exhibits.

13.2.4   Pertinent Microphone Specifications

The recording engineer needs to know the fundamental performance limits of a microphone as well as its general electrical characteristics.

The impedance of a microphone is a measure of its internal source impedance, and typical values seen today are in the range from 50 to 200 Ω. Such microphones are known as "low impedance" and constitute the bulk of professional microphones. Since microphones are normally operated into preamplifiers with input impedances in the range of 3000 Ω, there are rarely interface problems, and there is rarely a case of "loading down" the microphone.

The engineer needs to know the maximum sound field in which the microphone may be used. As a reference point, the sound pressure level that results in a total harmonic distortion (THD) figure of 0.5% is quoted for a given model.

At the other extreme, the engineer must be concerned with electrical input noise. In the case of capacitor microphones, the inherent self-noise generated by the microphone will be stated as an equivalent acoustical rating in dB(A). For example, a microphone with a noise floor of 15 dB(A) will generate an internal noise level equivalent to operating an ideal noiseless microphone in a sound field that has a noise floor of 15 dB(A).

Professional capacitor microphones have noise floors that vary from about 10 dB(A) up to perhaps 16 dB(A). Figure 13-6 shows the operating range of a high-quality capacitor microphone intended for critical studio applications.

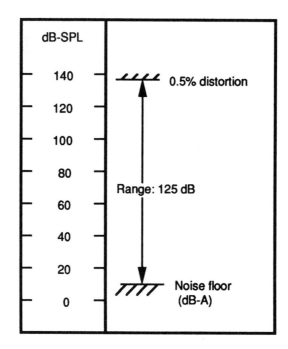

Figure 13-6. The effective dynamic range of a microphone.

13.3   BASIC STEREO PICKUP TECHNIQUES

Three relatively simple stereo pickup arrays are shown in figure 13-7. Figure 13-7a shows the so-called mid-side, or M-S, array, in which a figure-8 pattern is placed to pick up sound from the sides, while another pattern, usually cardioid, is used to pick up sound from the middle of the ensemble being recorded. These two signals are matrixed, or combined, in sum and difference networks to result in left and right outputs, as shown at the bottom of the figure.

The utility of the M-S system is that a recording made with it can easily be widened or narrowed in later stereo presentation by varying the amount of the S component in the final matrix combination.

The X-Y technique is shown in figure 13-7b. It consists of a pair of cardioid elements splayed at some angle, usually between 90° and 135°. Hypercardioid or supercardioid patterns may be used here as well. The amount of separation in the resulting pickup depends largely on the splay angle, and to a somewhat lesser extent on the specific microphone pattern.

The array shown in figure 13-7c was first described by Blumlein (1931) and is usually called simply the Blumlein array. Clark et al. (1956) modified the array slightly during the early 1950s, dubbing it the Stereosonic array. The array is equally sensitive in all azimuthal directions, and thus picks up a good bit of reverberant content if the recording space is live. The stereo imaging of sound sources between the frontal lobes is quite accurate and unambiguous. In studio application, sound sources may be placed behind the array for equally unambiguous pickup as well.

For obvious reasons the three microphone arrangements we have described are known as coincident arrays, and they depend solely on amplitude, or intensity, relationships to provide stereo localization in playback. Such recordings are often referred to as "intensity stereo."

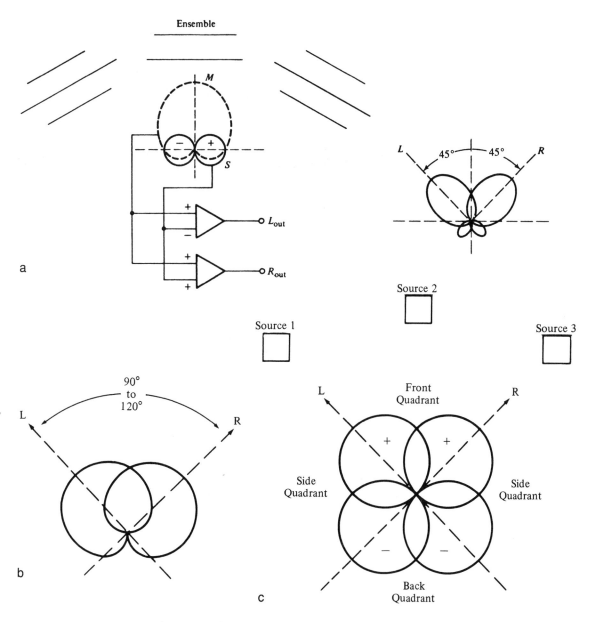

Figure 13-7. Coincident microphone arrays: (a) Mid-side; (b) X-Y; (c) Blumlein "crossed figure-8."

For those engineers who want time-related cues for playback localization, the arrays shown in figure 13-8 are useful. As a group, these are referred to as quasi-coincident or near-coincident arrays. (The Blumlein technique described in section 2.3.3 was the earliest of the quasi-coincident pickup methods, but it is rarely implemented as such today.)

The technique shown in figure 13-8a was introduced by the French Broadcasting Group (ORTF). It has many of the qualities of the Blumlein method, but is far easier to implement.

The technique shown in figure 13-8b uses a baffle to increase stereo separation at high frequencies, but allows some amount of signal commonality between the stereo channels at lower frequencies.

For recording engineers who wish enhanced spatial cues in recording, the various spaced-apart microphone techniques are useful. As a general rule, omnidirectional microphones are used for this, and they are spaced anywhere from 0.4 to 2 m (1.3 to 6.5 ft) apart, depending on the size of the group and the working distance. Such arrays are difficult to analyze, at least in a rigorous manner, but they often sound very good. What they lack in image specificity is often

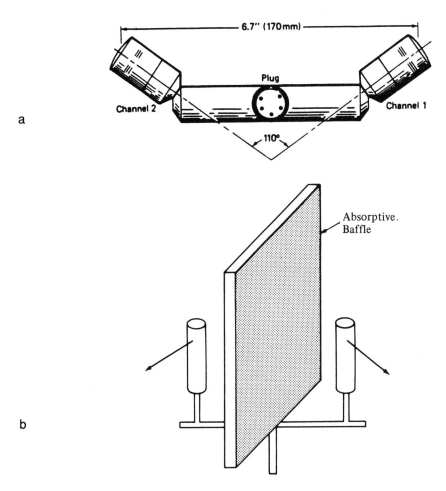

Figure 13-8. Quasi-coincident microphone arrays: (a) ORTF; (b) pair with baffle.

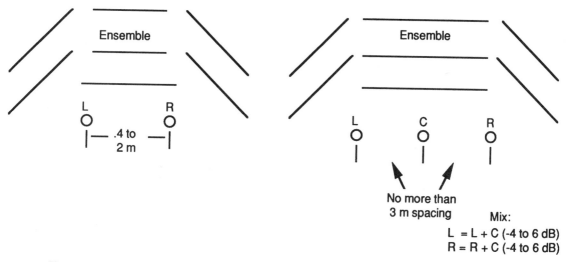

Figure 13-9. Spaced omnidirectional microphones for stereo.

made up for in other desirable qualities. Because of the microphone spacing, lateral time cues are enhanced, creating a feeling of spaciousness that may not truly exist in the recording environment. Localization is a result of both intensity cues and time cues, and as such may be a bit indistinct, especially in the middle of the stereo playback array. In order to counteract this, many engineers employ a center microphone, fed equally to the two stereo channels at a lower level, to "steer" center-stage events where they should be. When orchestral recordings are made with left-center-right spaced omnidirectional microphones, the overall spacing between adjacent microphones may be as great as 3 m (9.8 ft). There are no rules here, but the distance ratios shown in figure 13-9 are representative of current practice.

While spaced cardioid microphones may be used, most engineers who favor the spaced microphone approach will choose omnidirectional models, inasmuch as those microphones normally have more extended low-frequency response than do directional models.

From the radio broadcaster's point of view, coincident techniques are better than spaced-apart techniques in that monophonic compatibility is better. The spaced-apart arrays run some risk of producing signal cancellation when the stereo channels are summed for monophonic reproduction—and a great deal of FM radio listening is monophonic.

In the pop studio, there are few fundamental rules, and most pickup is through single microphones placed relatively close to individual instruments. The aim here is to achieve good separation so that an artificial stereo stage may be created later, during postproduction remixing.

13.3.1   Classical Recording Practice

In classical recording, most engineers and producers would agree that it is imperative to maintain a reference with the performance space. However, techniques can vary, and it is not unusual to see a combination of the methods we have described in this section. Figure 13-10 shows an approach to classical recording that crosses all barriers. The main pickup is provided by the ORTF array just behind the conductor. Flanking omnidirectional microphones are introduced into the stereo mix some 6 to 8 dB lower, just enough to add slight width to the strings, which are at the front of the orchestra, and to introduce added lateral time cues created by their rather large spacing.

The height of these microphones must be carefully chosen. As we saw in chapter 5, the polar distribution of energy from the violins is such that the engineer would prefer not to place the microphones along the normal axis of the instruments' top plates. Adjustments in height, as well as in fore and aft positioning, may be necessary to accomplish this. Furthermore, if the microphones are too high, there may not be the desired ratio of direct and reverberant sound. At this point, even small changes in positioning become significant, and there is no substitute for experience working in many venues.

The additional ORTF pair over the winds may be introduced into the mix 8 to 10 dB lower, and the purpose here is to add slight presence to those instruments. (Some engineers will time delay these microphones in order to have their signals arrive in step with the acoustically delayed signal. While this procedure is always correct, its audibility is dependent on thresholds that are determined by level differences and delay time settings.)

If the orchestra is recording on stage, it may be necessary to place a stereo pair of microphones in the house in order to adjust the balance between direct and reverberant sound. The reason is that the orchestra shell, although reflective, shields most of the orchestra from the reverberant field of the house. Stated another way, sound in the vicinity of the orchestra shell is largely influenced by direct sound from the instruments and first reflections from the shell. What is likely to be missing is room sound, and the house microphones will provide that.

If the orchestra is recording in a large studio, or in a ballroom converted for recording purposes, the extra reverberant pickup will probably not be necessary, inasmuch as the orchestra is placed well within the reverberant environment, and all microphones will pick up reverberant information along with direct sound.

Finally, a number of "accent" microphones may be deployed. These are used, hopefully subtly, to limn out lines or parts that would otherwise be lost in the stereo texture. Typically, the harp and celesta are provided with an accent microphone; the first stand of string basses may also be provided with one, just to give a bit more presence to that important line. Accent microphones may be delayed as well. The earliest description of the use of accent microphones was given by Maxfield (1947), so the technique is far from new.

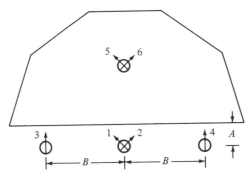

Figure 13-10. Basic setup for classical orchestral recording.

When recording in a traditional concert hall, many engineers prefer to build stage extensions in order to move the orchestra farther out into the acoustical space. Often, the back of the orchestra shell is moved forward to assist in this. It is not unusual for the audience seating area to be covered with reflective materials in order to liven the space.

13.3.2   Verisimilitude Versus Artistic License

The purist may be put off by the foregoing description of recording a symphony orchestra. To him, the event should simply be documented, and that is that! Presumably, this is done by strategically placing a stereo pair of microphones, and letting the orchestra and conductor take care of everything else. Unfortunately, events can rarely happen this way. In many recording venues, there is no such strategic microphone position that will work. More fundamentally, we must realize that stereo is a limited medium; there is no way that a single pair of channels can convey, at least over loudspeakers, the multiplicity of lateral reflections that one perceives in a concert hall. In effect, the recording engineer must "generate" the equivalent of this through thoughtful and intelligent usage of his tools, creating an array of stereo images and sound textures that are musically convincing over normal stereo loudspeakers. Add to this the further complications of recording under time and economic constraints, and in less than ideal venues, and one may begin to see what the recording engineer is really up against.

There are, fortunately, many smaller musical forms that can be successfully recorded with only two microphones. Recording engineers are doing themselves, and the music, a good service when they recognize these opportunities and take advantage of them.

13.4   MAGNETIC RECORDING

Today's recording practice has developed around the flexibility of multitrack tape recording and the convenience of editing various "takes" together to create the finished product.

Work in magnetic recording was done in the late 1800s, but its practicality in record production was not apparent until after World War II. The Germans had perfected the Magnetophone and had developed a tape medium capable of sufficiently low noise and distortion to enable it to replace disc recording. The change in the industry took place almost overnight.

Figure 13-11 shows the basic elements of an analog tape recorder. The tape is first passed over the erase head, where any remnant signals are erased through the action of a high-frequency bias signal. The tape then passes over the recording head, where a mixture of audio signal and high-frequency bias is applied to it. Bias has the effect of "linearizing" the medium, making it possible to record with quite low distortion over a wide dynamic range.

As the tape is being recorded, it can be simultaneously played back as a check on the reliability of the recording process. It can, of course, be edited, reassembled, replayed at a later time, and copied to other media.

The wavelength recorded on the tape determines the high-frequency response of the recording system. The wavelength must be longer than the playback head gap width to maintain response. If the wavelength is equal to the gap width, or a submultiple of it, then there will be nulls in the playback response. Typically, professional analog recording takes place at 38 cm/sec (15 ips), and the response extends comfortably beyond 20 kHz.

13.4.1   Characteristics of Magnetic Tape

Magnetic tape consists of a layer of gamma iron oxide (Fe_2O_3) on a plastic base. The oxide particles are acicular (needle-shaped) and are about 0.5 to 0.7 μ long. During the coating process, these particles are magnetically aligned longitudinally in the direction of eventual tape travel. They are thus sensitive to the applied magnetic flux and can change their magnetic state easily. Typical oxide thickness is about 12.5 μ.

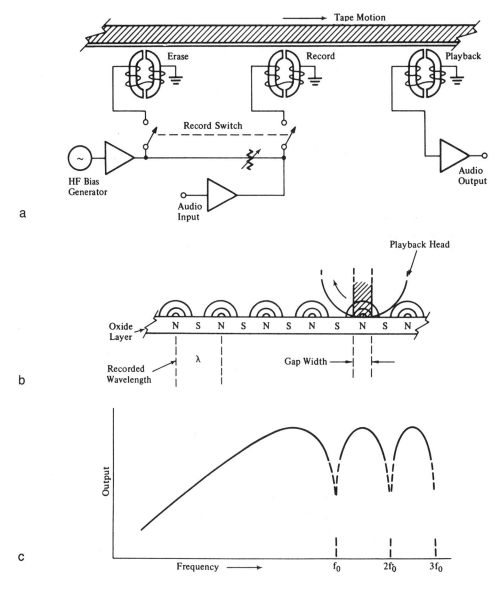

Figure 13-11. Principles of magnetic recording: (a) Head-tape configuration; (b) wavelength on tape; (c) wavelength-head relationships.

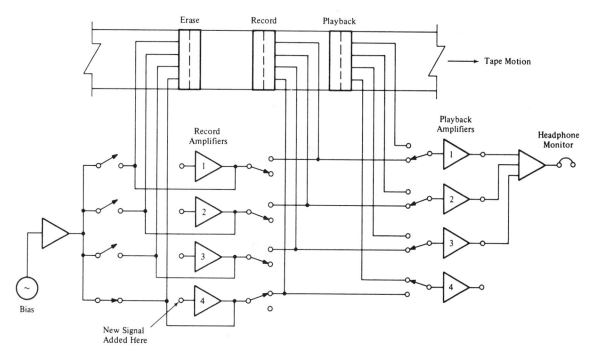

Figure 13-12. Principle of selective synchronization (Sel-Sync).

13.4.2 Capability for Overdubbing

Overdubbing refers to the practice of laying in new tracks over ones previously recorded. The technique is not new; in the days before magnetic recording, it could be accomplished by rerecording previous material, adding new parts in the process. This method resulted in at least one additional generation to the previous material, so the practice was not widely accepted.

In tape recording, a process known as Sel-Sync (trademark of the Ampex Corporation) can be used without increasing tape generations. The method is shown in figure 13-12. In this example, tracks 1, 2, and 3 have been previously recorded. The record heads for these tracks are used as playback heads and monitored by a performer, who then adds a new track (number 4) in synchronization with the other three.

Today's multitrack machines can accommodate up to twenty-four tracks. Not only can new tracks be added in synchronization with previous ones, but previously recorded tracks can be altered or corrected only where necessary by momentary Sel-Sync. Figure 13-13 shows a photograph of a typical studio twenty-four-track analog tape recorder.

The capability of overdubbing has fundamentally changed the creative process in the popular recording studio. Rhythm tracks may be recorded in one location, with vocals and other elements added later at a studio halfway around the world. A small group of strings may rerecord their parts more than once, creating the effect of more players. Various musical elements may be experimented with, and easily discarded if they do not work.

Figure 13-13. A modern multitrack tape recorder. Courtesy of Studer.

13.4.3   Digital Recording

Commercial digital recording was introduced in the early 1970s and has become the clear direction for future recording in the studio. The recording process is adaptable to tape, hard disc systems such as those used in computers, and optical disc media.

In analog tape recording, a magnetic replica of the audio signal is stored in a physical medium, and as such is subject to that medium's physical imperfections and inherent noise characteristics. By contrast, digital recording samples the audio signal at precise intervals in time and assigns a number that represents each sample's amplitude. It is the numbers that are stored, not the signal itself. As long as the numbers can be read back accurately, the original

signal can be recovered to the extent allowed by the sampling
numbers themselves. The numbers can be copied over many g
can be read accurately, the signal can be recovered with no s

The following analogy is useful. A third-generation Xerox
usually be correctly read, and thus contains exactly the sam
comparison, even the best photographic processes will show
original picture. The numbers (representing a digital reco
photographic process (representing analog recording) builds up cum.
succeeding copy.

The main parameters in a digital recording system are sampling rate and signal quantizing.
Many of today's digital recorders operate at a sampling rate of 44.1 kHz, and this sets the
absolute upper frequency limit of the system at 22.05 kHz, or half the sampling frequency. As a
matter of general practice, the system would be band-limited to about 20 kHz, just to ensure
an adequate frequency guard band. Sampling theory sets certain limits here, and it is necessary to have two samples per cycle in order to reconstruct the signal in its entirety.

Quantizing is the process of assigning a number to each sample. The binary number system is
used in digital recording, and it makes use only of two numerical values, 1 and 0. The following
list presents a comparison of binary and ordinary decimal counting. A single binary value is
referred to as a *bit*, while 8 bits may be referred to as a *byte*. In most digital recording systems,
quantization is carried out to 16 bits, and each quantized sample is referred to as a *word*.

Decimal Number	Binary Number	Decimal Number	Binary Number	Decimal Number	Binary Number
0	0000	6	0110	12	1100
1	0001	7	0111	13	1101
2	0010	8	1000	14	1110
3	0011	9	1001	15	1111
4	0100	10	1010		
5	0101	11	1011		

In this exercise we have counted from 0 to 15, a total of sixteen numbers. In the binary
system we require four columns of figures to do this. Note carefully that 2 raised to the fourth
power is equal to 16. This is the rule that determines the total of numbers available in the
system. If we consider a 16-bit system, we can calculate 2 to the sixteenth power, coming up
with the rather large value of 65,536.

What this states is that, in 16-bit quantization, we can subdivide the signal into 65,536
quantizing steps. We can determine the maximum signal-to-noise capability of the system by
noting that the maximum possible signal would exercise all possible values (65,536), while the
noise floor would be set by the least significant bit, which has a value of 1. We then take 20 log
(65,536), getting the value of 96.3 dB.

A simple rule of thumb is simply to multiply the number of bits of quantization by 6 to arrive
at the signal-to-noise ratio in decibels.

igure 13-14 illustrates the nature of the digital encoding process. For the sake of simplicity, e have shown a 5-bit system. An analog waveform (figure 13-14a) is quantized at time intervals that are the reciprocal of the sampling frequency, and this sets the resolution along the horizontal axis. Resolution along the vertical axis is set by the thirty-two values inherent in 5-bit quantization. The amplitude of the signal is quantized at the lowest binary value that just contains the analog signal. For each intersection of sampling and quantizing points, a 5-bit number is determined, and these numbers are stored by the recorder.

In any digital system, it is essential that the input bandwidth be limited to no more than one-half the sampling frequency. Any attempt to encode a higher frequency will result in cross-modulation products and distortion. It can be shown that a sampled system, digital or otherwise, can reproduce with absolute accuracy a sampled waveform provided that the frequencies present at the input are less than one-half the sampling frequency (Nyquist 1928).

When the signal is recovered in the playback operation, we still have the discrete, stepwise values shown in figure 13-14b. These are further filtered to produce the signal shown in figure 13-14c.

There are two processes that help the overall performance of the digital recording and playback system, filtering and dithering.

The analog input is filtered to remove components that are above one-half the sampling frequency. On playback, the signal is reconstructed through an additional filtering operation.

The low-level amplitude behavior of the system is improved by adding a small amount of random noise to the recording input. This is called *dither* noise, and the amount is carefully adjusted to be ±1/2 the least significant bit, which is just at the input threshold of the system. The purpose of dither is to make the system responsive to signal inputs that are in the range below the least significant bit of the system, or below about −96 dB.

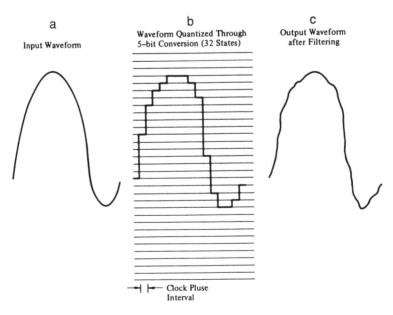

Figure 13-14. Digital sampling and quantizing: (*a*) Analog input; (*b*) the signal quantized and sampled in time; (*c*) output smoothing.

It is known that the ear can hear sine waves that are well below the broadband noise level of a transmission system. In the midband, such signals can be heard some 12 to 15 dB below the level of the noise floor, and the purpose of dither noise is to make the digital recording system responsive to these signals. Without dither noise, such inputs would virtually disappear in the quantization process. In a sense, they would "fall between the cracks."

Another benefit of dither noise is that it improves the recording system's distortion performance for very low-level input signals. This is a consequence of the fact that the dither noise makes the system effectively "continuous," rather than discrete, for low-level signals. Stated differently, it prevents any input signal from "sitting on the fence" between two quantizing levels.

In sum, a properly engineered digital recording system can preserve signal integrity from input to output to a degree determined only by its sampling rate and its quantization word length.

Digital recording systems are difficult to implement because the bit density on the recording medium is normally quite high. Elaborate error correction methods are employed to identify and correct readout errors resulting from dropouts, dust, and other momentary disturbances during playback, and these can vastly increase the complexity of both the recording and playback processes. Typically, the total requirement per channel might be as high as 1,000,000 bits per second.

A simplified block diagram of a digital recording system is shown in figure 13-15. The input signal is filtered, and dither noise is added to it. It is then fed to the analog-to-digital converter for the quantization process. From that stage it is formatted with synchronization and error correction information, then finally modulated onto the medium. Note that all steps have been under the control of a master clock.

On playback, the signal from the tape is stored in a buffer and clocked out as the error detection and correction system can handle it. The conditioned digital information is then fed to the digital-to-analog converter, and from there to the output reconstruction filter. As in the case of the recording operation, the process is timed by a master clock.

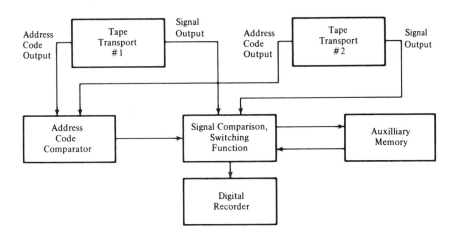

Figure 13-15. Simplified block diagram of digital recording and playback.

13.5   SIGNAL PROCESSING

From the earliest days of recording, engineers have found it necessary to condition the signal to fit the medium. This has taken many forms, including limiting its dynamic range as well as containing its frequency extremes.

In a more creative context, artificial reverberation may be added to recordings that may have insufficient natural reverberation. In popular recording, instruments may be brought to prominence through *equalization*, the emphasis of various parts of their frequency spectra.

Old recordings may be salvaged through judicious filtering of undesirable signal components such as hiss, hum, rumble, and noise.

These are among the recording processes referred to as *signal processing*. We will discuss them in the following order:

1. Dynamic range manipulation: compressors and limiters.

2. Frequency domain manipulation: equalizers and filters.

3. Time domain manipulation: reverberation and time delay.

4. Various special effects.

13.5.1   Limiters and Compressors

The input-output characteristics of limiters and compressors are normally shown by the convention indicated in figure 13-16. A linear amplifier is represented by a diagonal line, at 45°, running from lower left to upper right. For each decibel increase or decrease in input, the output matches it accordingly.

A compressor is a variable gain amplifier. At low levels, the output tracks the input; however, when the threshold of compression is reached, the output is reduced in level as the input increases. The compression ratio is defined as the ratio, measured in decibels, between an increase in input and the increase in output. For example, a compression ratio of 4 to 1 means that a 4-dB increase in input results in an output increase of only 1 dB. A limiter is a special compressor with a very high compression ratio, typically 10 to 1. Such a device effectively imposes a ceiling on the signal level passing through it.

Compressors are also characterized by their attack and recovery times. The attack time is the interval during which the sensing circuitry determines and implements the necessary gain reduction. The recovery time is the time interval required for the device to return to its unity-gain condition after the input signal has returned to normal levels. For compressors, attack times are normally in the range of tens of milliseconds, and recovery times are normally adjustable in the range of a handful of seconds, in order to fit specific musical requirements. For limiters, attack times and recovery times are typically much shorter.

Compressors and limiters are used in the following ways:

1. To contain modulation levels within safe limits for various media, such as broadcasting, tapes, and long-playing records.

2. To match various input signals to the natural dynamic range limitations of home music environments. These limits are the result of local ambient noise levels and the requirement of not disturbing one's neighbors.

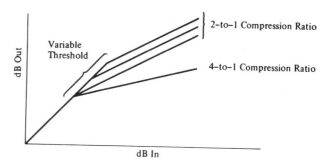

Figure 13-16. Input-output characteristics of a compressor.

3. To produce special effects. In popular or rock recording, many input signals can be transformed through judicious use of compression or limiting.

Two related devices are the noise gate and the "de-esser." The noise gate is a signal expander. That is, when the input signal drops below a certain threshold, the gain of the expander is reduced, diminishing unwanted input noise. The de-esser is a limiter that acts preferentially on high frequencies in the signal, thus removing troublesome sibilants from the program material. It is especially useful in broadcasting. While the de-esser is primarily a remedial tool, the noise gate, if its operating parameters are properly set, can be used to enhance certain signals by increasing the apparent dynamic range.

13.5.2   Equalizers and Filters

A transmission system is said to be "flat" if its transfer characteristic with respect to frequency is constant. The frequency response is represented by a straight, or flat, line on a graph of level versus frequency.

This is shown in figure 13-17a, along with various settings of a program equalizer. The term *equalization* dates from early telephony, when such devices were used to compensate for line losses, making the line output "equal" to the input. Today, we use the term to apply to any operation involving the alteration of frequency response for either remedial or creative purposes.

Filtering is a term applied to the removal of specific portions of the frequency spectrum. Such terms as low-pass, high-pass, bandpass, and band-reject are self-evident.

Various filter characteristics are shown in figure 13-17b.

Creative uses of equalizers and filters include:

1. Correcting spectral balance problems that occur as a result of close microphone placement. This usually entails reducing low-frequency rise resulting from the proximity effect.

2. Delineating specific instruments or musical lines by increasing high-frequency content.

3. Creating special audio effects, such as telephone transmission, short-wave transmission, and the like.

4. Altering the color, or spectral character, of a given instrument for purely musical reasons.

Remedial uses of equalizers and filters include:

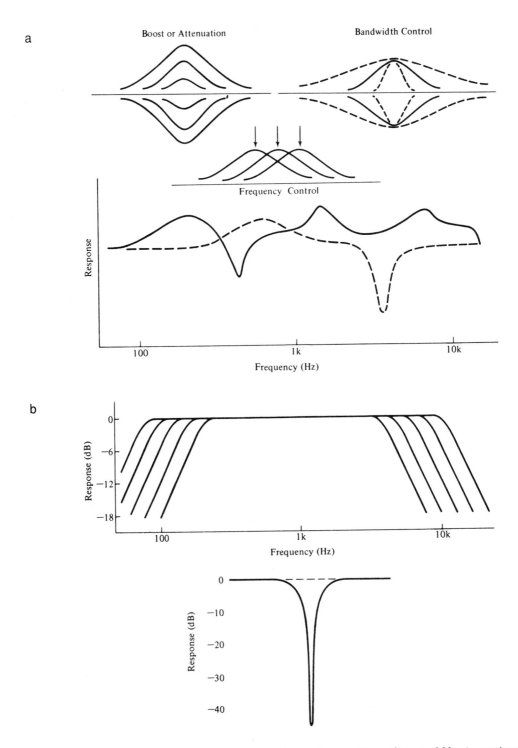

Figure 13-17. Equalization and filtering: (*a*) Typical equalizer settings; (*b*) typical filtering actions.

1. Removal of noise in various parts of the audio spectrum.

2. Correcting cumulative losses in the final transfer of older program material. Here, a trained ear is essential.

3. Correcting for various microphone characteristics. An example would be the use of a flat random incidence microphone in an application requiring flat on-axis response.

4. Shaping of program spectrum for transfer into difficult media, such as the cassette and the stereo LP record.

Today's recording consoles are liberally supplied with equalizers and filters in each input channel. A well-equipped studio will have a variety of outboard gear as well.

13.5.3   Signal Delay and Reverberation

Digital technology has brought the benefits of discrete time delay and reverberation generation into the recording studio, and at reasonable prices. The time-delay units provide upwards of a quarter second delay, often with resolution measured in tens of microseconds. The reverberation units accept stereo input and create either stereo or four-channel output. Among the variables that can be controlled are initial delay gap, low-frequency reverberation time, mid-frequency reverberation time, and high-frequency rolloff characteristic. Some models are calibrated in terms of volume of the simulated space, thus giving control over modal distribution and density.

All in all, today's better reverberation units are capable of natural, uncolored simulation of the real thing. For many studio applications, they can successfully simulate the natural sound field that might be picked up by microphones in a large and live environment. However, for the demands of classical recording, they are best used sparingly.

Figure 13-18 shows the use of time delay and artificial reverberation in creating a four-channel (quadraphonic) sound field.

Other uses for delay and artificial reverberation include the following:

1. Creation of early lateral reflections (normally accomplished by cross-feeding delayed signals from left to right and right to left).

2. Implementing the precedence effect (see section 2.3.2).

3. Implementing various image broadening and pseudostereo techniques (Eargle 1986a; Gardner 1973).

13.5.4   Noise Reduction

Noise reduction techniques have been successfully used in the studio since the 1960s. In principle, noise reduction systems employ signal compression during the recording function, thus enabling the recorded signal to remain well above the tape noise. On playback, the signal is expanded in a complementary manner, and the original program is virtually recovered. The various systems differ in technical detail, and they certainly differ in the audibility of their action. The Dolby A-type was the first professional system to be broadly accepted, and it has only been replaced by the Dolby SR (Spectral Recording) technique during the late 1980s. Figure 13-19 shows the basic action of a noise reduction system.

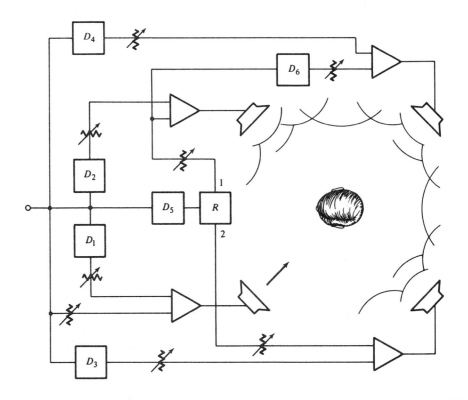

Figure 13-18. Signal flow diagram for producing a quadraphonic (four-channel) sound field with reverberation and time delay.

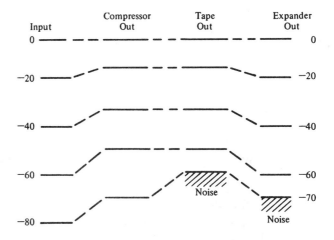

Figure 13-19. Basic compression-expansion action of a noise reduction system.

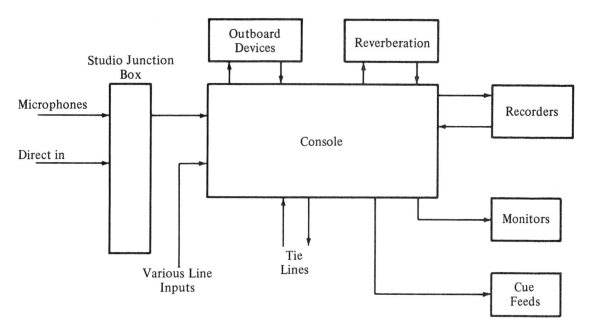

Figure 13-20. Basic console architecture.

13.6 RECORDING SYSTEM ARCHITECTURE

While modern recording consoles appear to be complicated, the basic signal flow through them is ordinarily quite simple and straightforward. This is especially true for classical recording, where a direct stereo (two-channel) product is desired.

In the modern pop-rock studio, things are basically no more complicated, but the demands for flexibility require far more switching capability so that the console can be reconfigured for various operations. Multitrack recording was developed primarily for these applications, and the flexibility offered by multichannel techniques allows engineers and recording producers to add tracks at a later date, correct or modify a segment of a previous track, or omit a track completely.

Most studio consoles today are of the *in-line* type. This approach facilitates the "one microphone, one track" recording philosophy, in which each musical ingredient is recorded on a single track. While this is being done, the signals can be experimentally mixed, equalized, reverberated, and so forth, so that producer, engineer, and artist can all hear a semblance of the finished product.

The remixing of the final product normally takes place later, using the same console, with special attention paid to the fine points of the mix, and without the pressures of a studio full of musicians.

A recording console today is conceived of as the control center for a vast number of operations, rather like the microprocessor at the heart of a computer. The simplified block diagram of figure 13-20 gives an indication of this.

REFERENCES

Blumlein, A. D. 1958. British Patent Specification 394,325 (1931) (Direction Effect in Sound Systems). J. *Audio Engineering Society* 6:91–98.

Bore, G. 1978. *Microphones*. Berlin: Georg Neumann Gmbh.

———, and S. Temmer. 1958. "M-S Stereophony and Compatibility." *Audio Magazine*.

Borwick, J. 1987. *Sound Recording Practice*. London: Oxford University Press.

Camras, M. 1989. *Magnetic Recording Handbook*. New York: Van Nostrand Reinhold.

Ceoen, C. 1972. "Comparative Stereophonic Listening Tests." J. *Audio Engineering Society* 20: 19–27.

Clark, H., et al. 1956. "The 'Stereosonic' Recording and Reproducing System." J. *Audio Engineering Society* 6:102–133.

Culshaw, J. 1967. *Ring Resounding*. New York: Viking Press.

Dolby, R. 1967. "An Audio Noise Reduction System." J. *Audio Engineering Society* 15(4).

———. 1987. "The Spectral Recording Process," J. *Audio Engineering Society* 35(3).

Eargle, J. 1982. *The Microphone Handbook*. Commack, NY: Elar Publishing Co.

———. 1986a. *Handbook of Recording Engineering*. New York: Van Nostrand Reinhold.

———. 1986b. "An Overview of Stereo Recording Techniques for Popular Recording." J. *Audio Engineering Society* 34(6).

Gardner, M. 1973. "Some Single- and Multiple-Source Localization Effects." J. *Audio Engineering Society* 21:430–437.

Journal of the Audio Engineering Society. 1986a. *Microphones*. New York: Journal of the Audio Engineering Society.

———. 1986b. *Stereophonic Techniques*. New York: Journal of the Audio Engineering Society.

Lipschitz, S. 1986. "Stereo Microphones Techniques—Are the Purists Wrong?" J. *Audio Engineering Society* 34(9).

Maxfield, J. 1947. "Liveness in Broadcasting." *The Western Electric Oscillator*.

———, and H. C. Harrison. 1926. "Methods of High Quality Recording and Reproducing of Music and Speech Based on Telephone Research." *Transactions of the* AIEE. Reprinted in O. Read and W. L. Welch, *From Tinfoil to Stereo*. Indianapolis: H. Sams, 1976.

Nyquist, H. 1928. "Certain Topics in Telegraph Transmission Theory." *Transactions of the* AIEE 47(2).

Pohlmann, K. 1989. *Principles of Digital Audio*. Indianapolis: H. Sams.

Snow, W. 1953. "Basic Principles of Stereophonic Sound." J. *Society of Motion Picture and Television Engineers* 61.

Steinberg, J. C., and W. B. Snow. 1934. "Auditory Perspective—Physical Factors." *Electrical Engineering* 53(1). Reprinted in *Stereophonic Techniques*, New York: Audio Engineering Society, 1986.

Woram, J. 1989. *Sound Recording Handbook*. Indianapolis: H. Sams.

————, and A. Kefauver. 1989. *The New Recording Studio Handbook*. Commack, NY: Elar Publishing Co.

14

High Fidelity Sound in the Home

One of the great successes of modern consumer electronics is the ready availability of high quality audio program presentation in the home. Evolution has been going on for about a century, beginning with the acoustical phonograph and progressing through radio, electrical recording, the long-playing record, television (both broadcast and via recordings), and stereophonic sound.

Today, the audio art has taken the lead with the digital compact disc. Television remains essentially as it was in the mid-1950s, and high-definition television (HDTV) is only now under development. Multichannel sound (more than two channels) has had a spotty history in the home as a purely audio medium; however, as an adjunct to video presentation, certain forms of multichannel sound are quite successful in the home environment.

14.1 BASIC COMPONENTS OF A HIGH FIDELITY SYSTEM

The essential components of a high fidelity system are:

1. Program sources

2. Signal control

3. Recording and processing loops

4. Amplifier-loudspeaker chain

These elements are shown in block form in figure 14-1a. Today, many of these functions are incorporated in a *receiver*, a device that incorporates a tuner (AM and FM), phonograph preamplification, signal control, and the power amplifier section, as shown in figure 14-1b.

In the following sections, we will discuss the principles by which the various signal sources operate. These will include AM radio, FM radio, the LP record, the analog tape cassette, and the compact disc (CD).

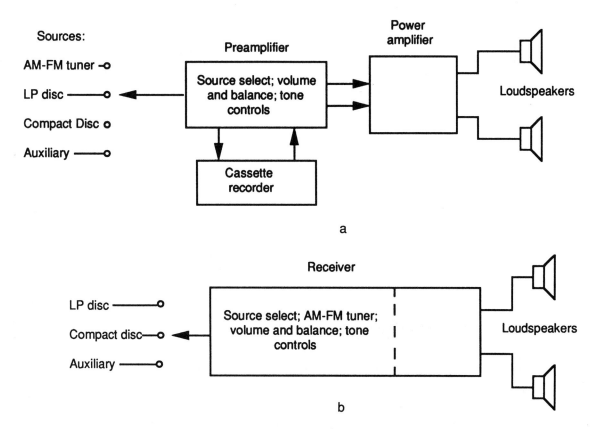

Figure 14-1. Basic high fidelity systems: (*a*) A system built around separate components; (*b*) a system built around a receiver.

14.2   AM RADIO TRANSMISSION AND RECEPTION

Commercial radio broadcasting dates from the second decade of this century, but the technique had been demonstrated by Guglielmo Marconi (1874–1937) in the late 1800s. AM is the abbreviation for *amplitude modulation*. In this method of transmission, the program signal is applied to a radio-frequency (RF) carrier signal, as shown in figure 14-2*a*. The amplitude of the carrier frequency is modulated by the program, as shown. In commercial broadcasting, carrier frequencies may be assigned in the range from 550 to 1600 kHz. Assignments of carrier frequency are made by government regulatory agencies in all countries, and adjacent broadcast facilities are normally spaced 20 kHz apart to avoid possible interference.

At the receiving end, a *tuner* locks in on a specific carrier frequency and amplifies it. The actual signal picked up by the antenna may be in the range of 10^{-15} W, so considerable amplification is required. The amplified signal is then fed to a *detector*, which rectifies the modulated carrier, producing a signal with a low-frequency component that is a replica of the program input. This signal is further filtered, amplified, and sent on to the loudspeaker. The receiving process is shown in figure 14-2*b*.

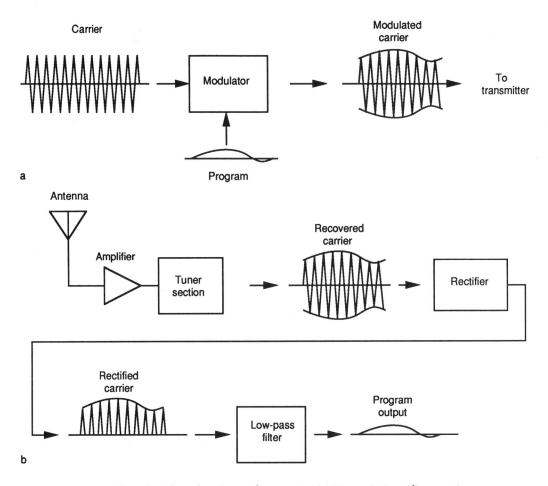

Figure 14-2. Principles of AM broadcasting and reception: (*a*) Transmission; (*b*) reception.

14.3   FM RADIO TRANSMISSION AND RECEPTION

AM radio reception is plagued by atmospheric electrical noises, as these disturbances produce amplitude-related disturbances in the carrier. While AM is still an important broadcast medium, it has largely been supplanted by FM for high-quality music transmission. FM is the abbreviation for *frequency modulation*. In this transmission method, the carrier is modulated up and down in frequency by the program source. Its amplitude remains constant and is thus not affected by atmospheric electrical disturbances. The basic principle is shown in figure 14-3*a* and *b*.

Since the frequency spectrum of FM is considerably wider than that of AM, an extra subcarrier can be added, enabling the medium to accommodate stereo. In this mode of transmission, the stereo channels, L and R, are summed to form a compatible baseband, or monophonic signal for reception on nonstereo receivers and tuners. When the subcarrier is demodulated, a difference signal is recovered that enables the following signals to be recovered:

$$\text{Baseband signal} = L + R$$

$$\text{Subcarrier signal} = L - R$$

Thus:

$$\text{Baseband signal plus subcarrier signal} = 0.5[(L + R) + (L - R)] = L$$

$$\text{Baseband signal minus subcarrier signal} = 0.5[(L + R) - (L - R)] = R$$

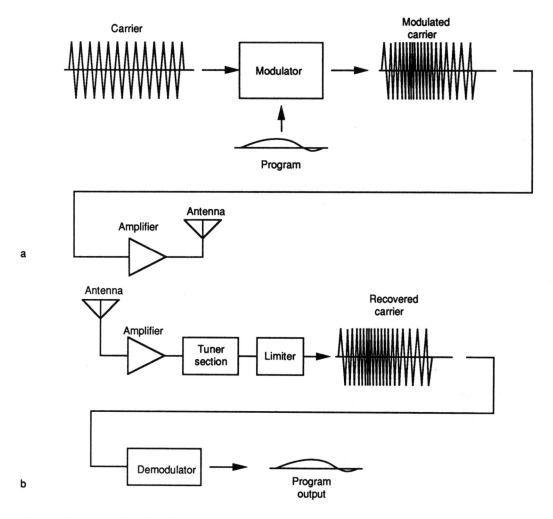

Figure 14-3. Principles of FM broadcasting and reception: (*a*) Transmission; (*b*) reception.

14.4 THE LONG-PLAYING (LP) RECORD

The LP record was the outgrowth of years of evolution of the disc medium. It was introduced as a monophonic (single-channel) medium in 1947, and in 1958 the stereophonic LP was introduced. Since 1983, the LP has been eclipsed in sales by the analog tape cassette, and since 1988, the compact disc has overtaken it in the marketplace. While the demise of the LP has long been predicted, it will be with us for some time to come, primarily as a result of its long and rich history of musical documentation. In time, the great performances of the past will find their way onto compact disc, and the LP may be retired. For now, however, it remains an important medium.

The basic characteristics of the LP are:

1. Playing time per side: upwards of 30 min. depending on program

2. Frequency response: 20 Hz to 20 kHz, depending on recording diameter

3. Signal-to-noise ratio: about 60 dB

Figure 14-4a shows a scanning electron microscope view of stereophonic record grooves. The left channel signal is cut into one groove wall, and the right channel signal is cut into the other. Section views of the groove are shown in figure 14-4b through e. Since the motions of the two groove walls are at right angles to each other, a high degree of interchannel separation can be maintained.

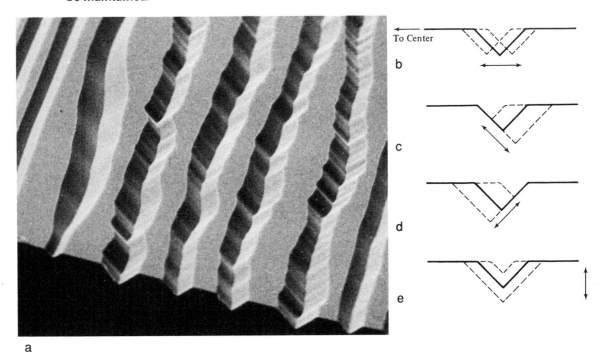

Figure 14-4. The long-playing record. (a) Scanning electron microscope view of grooves; (b) through (e) section views of groove for various signals: (b) lateral (monophonic) motion; (c) stereo right channel only; (d) stereo left channel only; (e) vertical motion. Photo courtesy JVC Corporation.

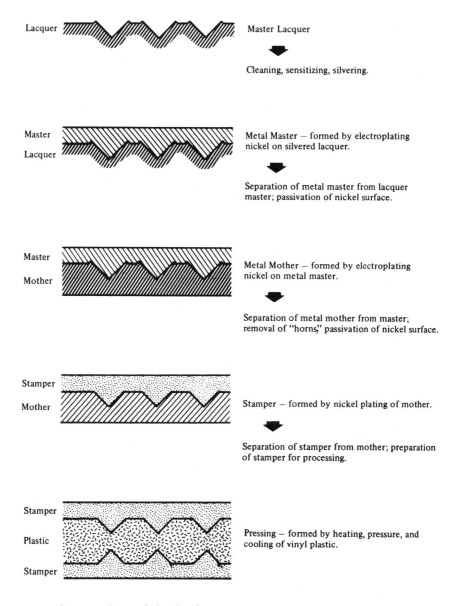

Lacquer — Master Lacquer

Cleaning, sensitizing, silvering.

Master Lacquer — Metal Master — formed by electroplating nickel on silvered lacquer.

Separation of metal master from lacquer master; passivation of nickel surface.

Master Mother — Metal Mother — formed by electroplating nickel on metal master.

Separation of metal mother from master; removal of "horns," passivation of nickel surface.

Stamper Mother — Stamper — formed by nickel plating of mother.

Separation of stamper from mother; preparation of stamper for processing.

Stamper Plastic Stamper — Pressing — formed by heating, pressure, and cooling of vinyl plastic.

Figure 14-5. Replication of records by the three-step process.

Records are replicated by plating the master electrolytically, and subsequently replating to produce stampers, which are negatives of the original. The stampers are then placed in a press, and heated plastic is forced under compression to conform to the shape of the stampers. The records are cooled before the press is opened, and the finished product is trimmed, inspected, and packaged. The basic manufacturing process is shown in figure 14-5.

14.5   THE CASSETTE

The cassette analog tape medium was developed by Philips and introduced during the 1960s. After many refinements in tape formulation, magnetic head design, mechanical design of transports, and application of noise reduction, the cassette has emerged as a high-quality storage medium. It effectively replaced the reel-to-reel tape recorder in the home during the 1970s.

Since the tape is entirely contained within the cassette, there are no threading problems for the user. Tape speed is 4.76 cm/sec (1.875 in./sec), and tape width is about 3.5 mm. There are four tracks available. In the normal home format, they are utilized as two stereo pairs, with the tape recorded in both directions, as shown in figure 14-6. The principle of operation is as discussed in section 13.4.

As a result of developments in the medium, there are several recording and playback standards, so the consumer must pick the correct bias and equalization settings before recording. At the present time, there are no fewer than four noise reduction standards that have been used with the cassette. The Dolby B and C formats have been dominant in the area of consumer noise reduction, but the newer Dolby S format is likely to replace them in the future.

14.6   THE COMPACT DISC (CD)

Jointly developed by Philips and Sony, the CD has enjoyed great success in just a few years. It was introduced in 1983, and since 1988 unit sales have exceeded those of the stereo LP. As a digital medium, it is virtually free of ticks, pops, and the myriad of other problems that have long plagued the analog LP. Its performance is based on the technology discussed in section 13.4.3, but it is a playback-only medium. Figure 14-7 shows a detail of the signal as it is molded into the disc. A series of pits reflect the beam from a laser in such a way that the digital signal is

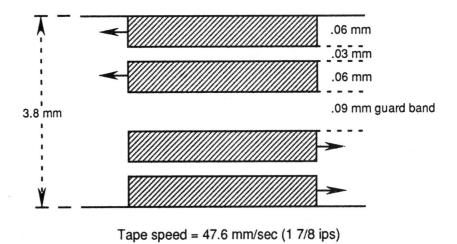

Tape speed = 47.6 mm/sec (1 7/8 ips)

Figure 14-6. Cassette track format.

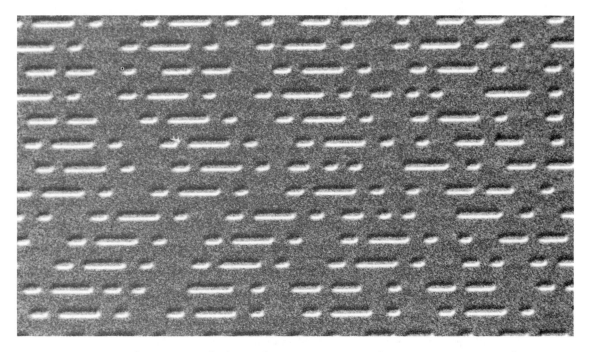

Figure 14-7. Photomicrograph of a CD showing pit structure. Photo courtesy University of Miami.

recovered. There is no mechanical contact, and the pits themselves are protected by a clear layer of plastic. The discs are virtually indestructible, and of course there is no wear. Economies of scale in both player manufacture and pressing have brought the cost of the medium down dramatically.

The CD has the following mechanical characteristics:

1. Linear track velocity: 1.2 m/sec

2. Disc diameter: 120 mm

3. Pit width: about 0.5 mm

4. Pit depth: about 0.11 mm

5. Distance between adjacent tracks: 1.6 mm

6. Pit length: 0.833–3.054 mm

The performance characteristics are:

1. Quantization: 16-bit linear

2. Sampling frequency: 44.1 kHz

3. Frequency response: flat to 20 kHz

4. Signal-to-noise ratio: greater than 90 dB

5. Channel capacity: two (four with reformatting)

14.7   SIGNAL CONTROL

The basic signal control functions in a high fidelity system are shown in figure 14-8. Volume and balance control are shown in figure 14-8*a*, and the recording loop is shown in figure 14-8*b*.

Typical tone control curves are shown in figure 14-9*a*, and loudness contours are shown in figure 14-9*b*.

14.8   AMPLIFIERS

An amplifier is an active device for increasing the levels of audio signals. We may speak of voltage amplifiers (those that increase voltage) or of power amplifiers (those that increase the power of the signal for driving a loudspeaker). The device known as a *preamplifier* is an assemblage of voltage amplifiers and associated circuitry for performing various switching or equalization functions. In contemporary design, the integrated circuit (IC) operational amplifier (op amp) is widely used for voltage amplification. It is shown in functional form in figure 14-10*a*. The gain of an operational amplifier is accurately determined by the ratio of the feedback and input resistances, R_f and R_i.

A functional diagram for a power amplifier is shown in figure 14-10*b*. A device such as this can, depending on the number and size of the devices in the driver and output stages, deliver power in the range of several hundreds of watts for driving loudspeaker loads.

Virtually all voltage and power amplifiers manufactured today are of the solid state type, using transistors rather than vacuum tubes. For the typical home high fidelity system, a stereo power amplifier capable of delivering 100 to 200 W per channel into 8-Ω loudspeakers would be appropriate.

14.9   LOUDSPEAKERS

Most loudspeakers used today are of the dynamic variety, deriving their acoustical output through a voice coil placed in a magnetic field. In that regard, they are the inverse of the dynamic microphone discussed in section 13.2.2. Less commonly, we find ribbon and electrostatic loudspeakers, both operating along principles similar to those of microphones.

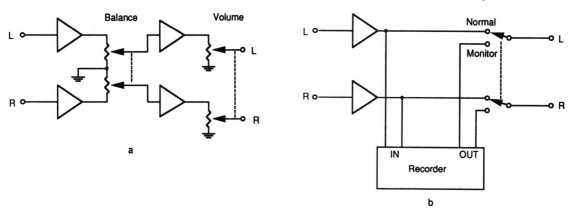

Figure 14-8. Control functions: (*a*) Volume and balance; (*b*) external recording loop.

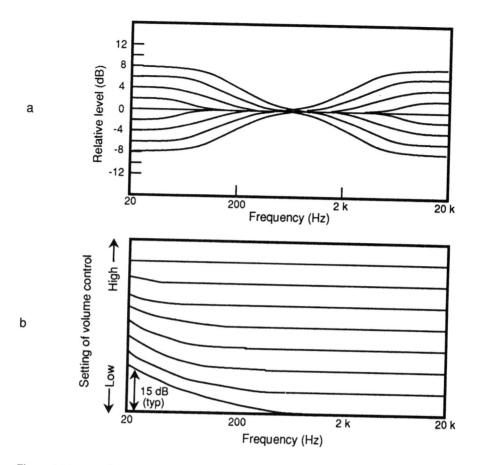

Figure 14-9. Equalization: (a) Typical tone control response curves; (b) loudness compensation.

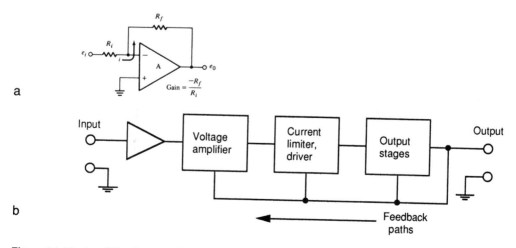

Figure 14-10. Amplification: (a) The operational amplifier; (b) a power amplifier.

The cone loudspeaker is basically of bandpass nature, and the natural range of a loud-speaker is bounded at low frequencies by the basic resonance of the cone, f_0, and at upper frequencies by the leveling off of its radiation resistance at f_1. Thus, the typical loudspeaker system will consist of several radiating elements, each optimized for its portion of the frequency spectrum.

A dividing network is used to channel the input signal appropriately into each driver in the system. Figure 14-11a shows the basic bandpass nature of a cone loudspeaker, while a complete system is shown in figure 14-11b.

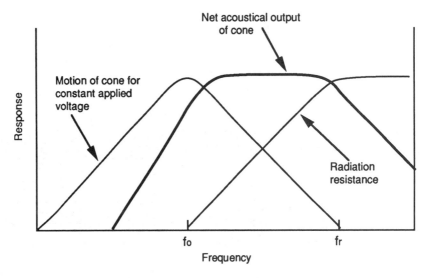

fo = resonance frequency of cone

fr = frequency above which radiation resistance is flat

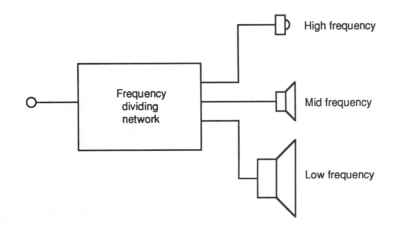

Figure 14-11. Loudspeakers: (a) Basic bandpass nature; (b) a complete loudspeaker system.

The low-frequency driver of the system may be mounted in either a sealed or a ported enclosure. Where greater low-frequency output is desired, the ported enclosure offers advantages in that the tuned enclosure resonance (its Helmholtz resonance) can offer added output capability with minimum cone excursion. On the other hand, the sealed enclosure may be capable of extending to lower frequencies, although with somewhat reduced output capability.

14.10   LISTENING ROOM CONSIDERATIONS

A good listening room should be fairly neutral; that is, it should not draw attention to itself through prominent reflections or room modes. Perhaps it is better if a room is too dead rather than too live, but in any event care should be taken that the room does not approach anechoic conditions.

The loudspeakers should ideally be placed so that they are symmetrical with some axis of the room, and the subtended listening angle for a listener located along the plane of symmetry should be in the range of 45° to 60°. Narrower placement, which some people prefer, tends to produce a subtle stereo sound stage, while wider placement creates instability in the reproduced stereo stage. Slight toe-in of the loudspeakers generally tends to widen the area of stereo perception by placing off-axis listeners more along the "firing line" of the more distant loudspeaker.

REFERENCES

Bech, S., and O. Pederson. 1987. *Perception of Reproduced Sound*. Denmark: Gammel Avernæs.

Borwick, J. 1988. *Loudspeaker and Headphone Handbook*. London: Butterworths.

Camras, M. 1988. *Magnetic Recording Handbook*. New York: Van Nostrand Reinhold.

Collums, M. 1978. *High Performance Loudspeakers*. London: Pentech Press.

Eargle, J. 1986. *Handbook of Recording Engineering*. New York: Van Nostrand Reinhold.

Harvith, J., and S. Harvith. 1987. *Edison, Musicians, and the Phonograph*. Westport, Conn.: Greenwood Press.

Journal of the Audio Engineering Society. 1986. *Stereophonic Techniques*. New York: Journal of the Audio Engineering Society.

Olson, H. 1952. *Musical Engineering*. New York: McGraw-Hill.

Pohlmann, K. 1989. *The Compact Disc*. Madison, WI: A-R Editions, Inc.

Read, O., and W. Welch. 1976. *From Tin Foil to Stereo*. Indianapolis: H. Sams.

Strong, W., and G. Plitnik. 1977. *Music, Speech, & High Fidelity*. Provo, UT: Brigham Young University Press.

15

Overview of Music Synthesis

Music synthesis generally refers to the creation of musical sounds by electrical (nonacoustical) means. Techniques for doing this have been known for many years. One of the earliest was the *telharmonium*, invented by Thaddeus Cahill during the first decade of this century. The device employed rotating toothed wheels that imparted their signals to nearby electromagnets. Various tones could be combined in harmonic relationships, and complex signals could be produced and transmitted over telephone lines.

Later, with the advent of electronics, oscillating circuits came into their own, and such instruments as the Theremin (1927) were invented (Léon Thérémin, b. 1896). Thérémin's instrument was played by moving the hand in the vicinity of a vertical antenna; this changed the electrical capacitance in an oscillating circuit, and thus the pitch of the tone. The other hand altered the volume of sound. The Theremin was capable, like an unfretted string instrument, of playing all pitches in its range, and considerable skill was required for truly musical results. Later, the *ondes martenot* (1928) (Maurice Martenot, b. 1898) provided a similar playing technique, coupled with a keyboard and a ribbon along which the player's thumb could slide to produce continuous pitches.

The 1930s saw the rise of the first electronic organs. That of Laurens Hammond (1895–1973), developed in 1933, was undoubtedly the most successful; its technology was based on that of the telharmonium. The Everette Orgatron (1935) made use of free reeds, whose oscillations were capacitively picked up and amplified. During the 1940s more sophisticated approaches were developed, and it became apparent that the art was moving beyond the basic level and proving itself capable of producing musical sounds.

The problems that remained were those of carefully analyzing the *musical envelope*, that is, the attack, growth, sustaining, and decay of musical sounds. Most of the early electronic instruments treated tones simply as steady-state signals that were abruptly turned on and off or, in the case of the Orgatron, allowed to rise slightly in pitch when the wind excitation was stopped—all of which sounded unnatural.

The acceptance of synthesized sound in recent years has been a natural consequence of the many improvements that have been made in envelope control and in natural musical factors such as pitch and amplitude randomization. Electronically generated sounds do not necessarily have to imitate those of traditional instruments, but the sounds do have to fit into musical contexts.

15.1   ANATOMY OF A MUSICAL SOUND

Musicians speak of such things as attack, timbre, and release when they describe isolated musical sound in a qualitative way. Many of these factors can be analyzed and implemented in the synthesis of musical sounds. We will discuss them in the following sections.

15.2   ATTACK: THE INITIAL TRANSIENT

When a tone is produced on a musical instrument, a resonating system is forced into vibration. Mechanical and acoustical inertance (mass) prevent the instrument from sounding immediately, and before the steady-state condition is reached, there will be a number of short-lived events that are intimately part of the sound of an instrument. Indeed, the effect of the starting transient in many complex musical settings is to assist the listeners in sorting out musical lines (Winckel 1967).

For example, the initial action of a bow on a violin string produces a scratch, which exists until the string is in oscillation. (To some extent, the scratch persists throughout the sounding of the string.) If a recording is made of a violin tone, and this initial transient is removed through editing, the tone will not necessarily sound like that of a violin, and many listeners will fail to identify its source.

In like manner, a flue organ pipe will exhibit a "chiff" before its tone reaches its steady state. The chiff is actually random noise produced by the jet of air against the upper lip of the pipe, and its composition may vary markedly from pipe to pipe of the same stop.

A piano tone is characterized by the low-frequency thump of the hammer; again, if such transient events are removed from recorded tones, the results are often identified as being "electronic sounding."

In addition to these transients, which are all nonharmonic in structure, there will be initial variations in the harmonic content of the tones. Figure 15-1a shows the basic envelope structure for a musical tone. Here, only a single harmonic has been shown. The representation

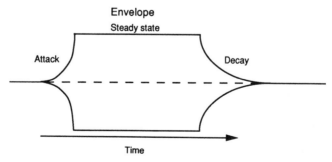

a

Figure 15-1. Musical attacks and starting transients. (a) The idealized attack–steady-state–decay envelope. Harmonic contributions in starting transients for (b) a flue organ pipe; (c) a trumpet; and (d) a violin. Data after Winckel 1967.

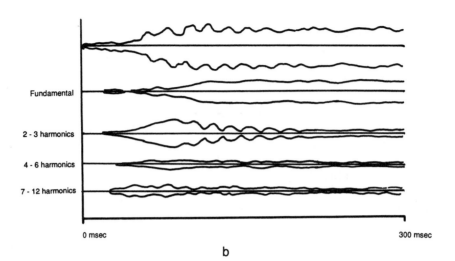

Fundamental

2 - 3 harmonics

4 - 6 harmonics

7 - 12 harmonics

0 msec   300 msec

b

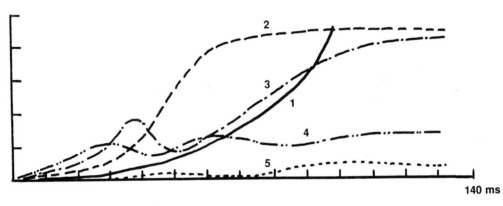

140 ms

c

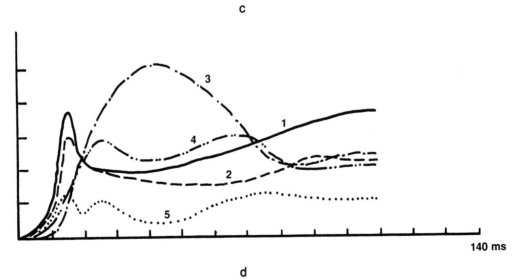

140 ms

d

in figure 15-1*b* shows various harmonics of a flue organ pipe. Note that the upper harmonics build up more quickly than the lower ones. A slight amount of inharmonicity is apparent in the undulation of the second and third harmonic content.

An example of a trumpet attack is shown in figure 15-1*c*, and a violin attack is shown in figure 15-1*d*. Note that the first and second harmonics of the trumpet are relatively quick to speak, but both fall immediately and then rise slowly. The violin produces a relatively slow overall attack with normal bowing.

If musical tones are to be synthesized, their onset transients must be analyzed, at least through the first five or six harmonics, and that analysis used to construct a natural-sounding attack. Nonharmonic contributions must be analyzed and similarly re-created. If new sounds are to be created, then the "designer" must invent a suitable starting transient and incorporate it into the plan.

15.3   STEADY-STATE CONDITIONS

For continuous tones, the starting transient is followed by a steady-state condition. One of the problems with traditional electrical oscillators is that they produce tones of complete regularity and predictability, no matter how complex those tones may be. In nature, no such regularity exists; there are several departures from it:

1. Inharmonicity. The harmonics in many musical resonators are not exact multiples of the fundamental. Figure 15-2 shows a steady tone produced on the trumpet (G_4), and it can be seen that some of the higher harmonics are wandering, as it were. These are apparent in that part of the waveform close to the baseline.

2. Amplitude and frequency randomness. It is difficult to maintain a steady, unwavering tone, and there will always be slight amplitude and pitch changes in any steady musical tone. Again, such variations are not a part of traditional electrical oscillators.

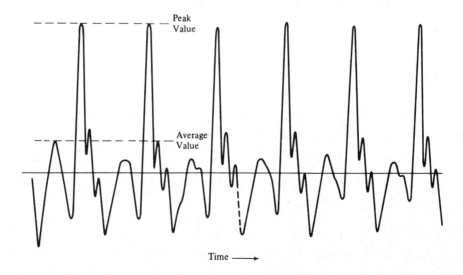

Figure 15-2. Steady state of a trumpet tone (G_4).

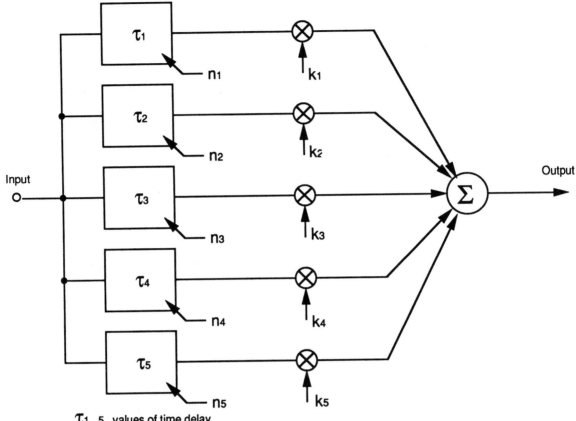

$\tau_{1..5}$, values of time delay

$n_{1..5}$, modulating values of low-frequency noise

$k_{1..5}$, modulating values of low-frequency noise

Figure 15-3. A method for stimulating a chorus effect from a single source by adding random timing and amplitude variations.

3. Vibrato. Many instruments are played with a vibrato. While this is essentially a slight variation in pitch in the 5- to 7-Hz range, it is usually accompanied by some degree of amplitude variation. Furthermore, the precise speed of a vibrato depends on many musical factors, such as instrumental range and the tempo of the music. Another important aspect of the vibrato is that it often entails spectral variations as well (Fletcher and Sanders 1967).

If traditional instruments are being synthesized, it is clear that all aspects of the steady-state performance must be identified. If new sounds are being invented, a suitable amount of both amplitude and frequency randomness, and perhaps a degree of inharmonicity, must be included in the synthesis scheme. While the ear disdains a perfectly regular waveform, it may not take a great deal of added randomness in these various areas to satisfy the ear.

15.4   DECAY OF TONES

For tones that are struck, there is no steady-state condition, and the tone begins to decay shortly after it is struck. In general, when excitation ceases, the upper harmonics damp out faster than the lower ones. In many cases, we are hardly aware of the precise nature of the decay of a note, since it can so easily be masked by another attack.

The tam-tam is an interesting case. While the starting transient for most musical sounds is over in a matter of half a second or less, that of the tam-tam, depending on how it is struck, unfolds over a period of a few seconds, with some partials decaying while others are still rising.

The decay of a piano note masks both the inharmonicity in individual strings and the slight degree of tuning error between them.

The synthesist must be aware of subtleties of decay and incorporate them as needed.

15.5   ENSEMBLE CONSIDERATIONS

Eventually, all synthesized tones must come together in some performance environment. They may be performed, or they may be stored and played back from a recording. In any event, the synthesist must be aware of the requirements for certain pitch and temporal "fringes."

Let us assume that the synthesist wishes to create the sound of a large symphonic violin section, or something essentially like it. In addition to the problems of creating the sound of a single violin, the synthesist now has to face the task of simulating a number of them coming from many directions. It probably is not necessary to use twenty-four discrete synthesized violins and present them in stereo. Very likely, it will suffice to synthesize only five or six violins, vary their relative timing slightly, vary their relative pitches slightly, and make sure that they are perceived as arriving from a multiplicity of directions.

The system shown in figure 15-3 produces an effective chorus effect through random shifting of both pitch and amplitude (Blesser and Kates 1978). A small number of these systems, fed with slightly "different" violins and presented from several locations, would suffice to synthesize a large number of violins.

15.6   REPRESENTATION OF STEADY-STATE WAVEFORMS IN TIME AND FREQUENCY DOMAINS

Depending on the tools at hand, the synthesist may work in either the frequency or the time domain in creating waveforms. Figure 15-4 shows the time representation of a number of waveforms and their corresponding frequency spectra.

15.7   THE PERFORMER'S INTERFACE

Today, the word *synthesizer* conjures up images of performers seated at keyboard instruments that can be programmed to sound like almost anything! The various keyboard-based synthesizers have entered the mainstream of modern popular and rock performance, and the worlds of motion picture and television scoring would be quite different without them.

But not too many years ago, the term represented a large, stationary rack of equipment, all accessed through an elaborate patching system and actuated by programmed commands.

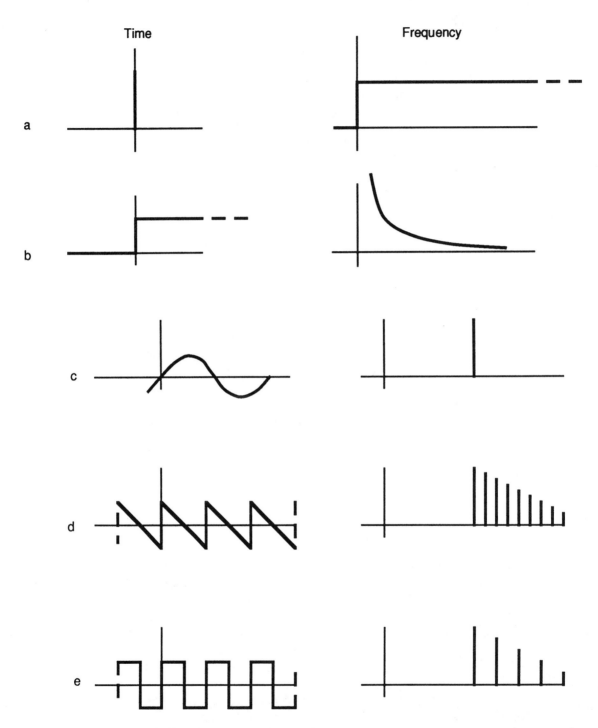

Figure 15-4. Time and frequency domain equivalents: (*a*) Unit impulse function; (*b*) step function; (*c*) sine wave; (*d*) sawtooth wave; (*e*) square wave.

While the electronic music departments on many university campuses have such installations, they represent a quite different application from real-time performance.

Some of the features that we take for granted in today's keyboard-based synthesizers are:

1. Sampling. Individual sounds may be sampled, recorded and stored digitally, and processed so that they can be played at various pitches directly from the keyboard. The technique was probably first used by some of the more adventurous builders of electronic organs who were looking for ways of making their instruments produce a true imitation of pipe organ tone.

2. MIDI (musical instrument digital interface). MIDI is a digital code for transmitting control and timing information between synthesizers in real time (Moog 1986). With it, compatible synthesizers can be played from a single keyboard, and all the commands possible with a given synthesizer can be accessed through the interface by another synthesizer. There are also pitch-to-MIDI converters that allow a conventional instrument, such as a guitar or piano, to generate a MIDI signal, which can in turn bring the synthesizers into play.

For the most part, digital technology has replaced analog methods, because it can be so easily programmed by appropriate system software. By comparison, analog equipment must be physically programmed through patching of inputs and outputs. Digital equipment is more stable, and it enjoys the benefits of increased memory and capability in ever decreasing package size.

As for the future of music synthesis, everything looks rosy indeed. Economic considerations alone will spur growth, since musicians' costs are ever increasing. And our ears are becoming even more conditioned and receptive to the sounds of these remarkable instruments.

REFERENCES

Blesser, B., and J. Kates. 1978. "Digital Processing in Audio Signals." In *Applications of Digital Signal Processing*, edited by A. Oppenheimer. Englewood Cliffs, NJ: Prentice-Hall.

Fletcher, H., and L. Sanders. 1967. "Quality of Violin Vibrato Tones." J. *Acoustical Society of America* 41:1534–1544.

Moog, R. 1986. "MIDI: Musical Instrument Digital Interface." J. *Audio Engineering Society* 34(5).

Rossing, T. 1990. *The Science of Sound*. Reading, MA: Addison-Wesley.

Strong, W., and G. Plitnik. 1977. *Music, Speech & High Fidelity*. Provo, UT: Brigham Young University Press.

Winckel, F. 1967. *Music, Sound, and Sensation: A Modern Exposition*. New York: Dover Publications.

16

Active Noise Control

Traditionally, the acoustical engineering community has taken the point of view that it is better to avoid noise problems at the design stage than to eliminate them later. When problems do occur, the traditional, and probably the most straightforward, way of solving them is through the application of passive methods, such as acoustical absorption, isolation mounts, and the like. These methods are in fact the only sensible way of combating broadband noise in large public assembly spaces.

However, in many industrial environments, very high noise may be encountered, much of it at low frequencies. Passive absorbers that work effectively at low frequencies are quite large and often do not represent an acceptable solution to the problem.

The principle of active noise control is as old as negative feedback in electronic amplification and servomechanism control systems. Noise, after all, is an unwanted signal. If it can be identified at its origin, it can be amplified, processed, and returned in inverse polarity, thus canceling itself.

This idea can also be extended to the problems of acoustical damping, where making a space less responsive to locally generated sounds may be desirable. Here, we are not referring to noise reduction per se, but to "assisting" the normal room boundaries in the absorption of sounds originating in that space.

Active methods can be of benefit in the following areas:

1. Reduction of localized sound pressure fields in the neighborhood of the listener or worker. This application may be of great importance for hearing conservation in noisy and hazardous environments.

2. Increasing acoustical absorption at room boundaries. This technique has great application in many critical listening environments, where the passive alternative may require extensive alteration of the physical space.

3. Reduction of low-frequency noise originating in connection with rotating machinery, air turbulence in industrial exhaust systems, and similar situations where the disturbing noise source is localized. In such cases, the benefit in noise reduction may extend over a large area.

In this chapter we will examine these applications.

16.1   PRINCIPLE OF ACTIVE SOUND CANCELLATION

Olson and May (1953) described a practical electronic sound pressure cancellation system. The basic element in this system was a microphone–amplifier–loudspeaker combination, as shown in figure 16-1a. The amplifier inverts the signal, providing an "out-of-phase" relationship between its input and output such that a pressure p_1 appearing at the microphone gives rise to a negative pressure p_2 produced by the loudspeaker, which tends to cancel p_1.

The effectiveness of the system depends on its gain. In order for the system to be operated at high gain, it must be stable under all operating conditions, and this requires microphones and loudspeakers that have extremely smooth low-frequency response.

Many capacitor omnidirectional microphones provide such response and are available at very reasonable cost. The loudspeakers are another matter. They must be capable of very linear response and must operate at a low resonance frequency while mounted in a sealed enclosure of modest size. The electromagnetic damping in the loudspeaker should be fairly high, and this requires a sizable magnet; such loudspeakers are not cheap. The response curves shown in figure 16-1b and c show the response achieved by Olson in his developmental work.

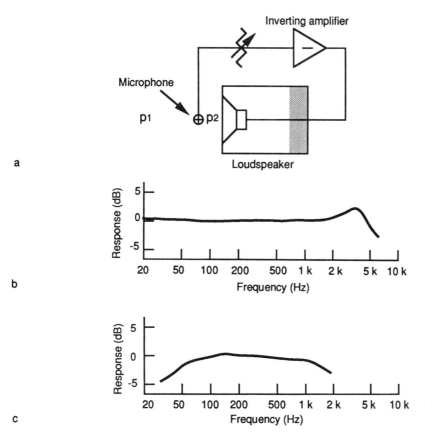

Figure 16-1. Principle of sound pressure cancellation. (a) Basic pressure-canceling module; (b) microphone response; (c) loudspeaker response. After Olson and May 1953.

16.2   SOUND PRESSURE REDUCTION

The range of operation of Olson's basic system is shown in figure 16-2*a*. The measurement points were at various distances from the microphone (figure 16-2*b*). Up to 24 dB of pressure cancellation was attained in the 50- to 100-Hz range at the face of the microphone. For positions farther away from the microphone, the pressure cancellation was correspondingly less, since the spherical waves from the pressure cancellation module are losing level as a result of inverse square relationships. Vector relationships are shown in figure 16-2*c*. Here, p_2 and p_1 represent vectors that sum in space, creating a resultant vector, p_r, that is 25 dB lower in

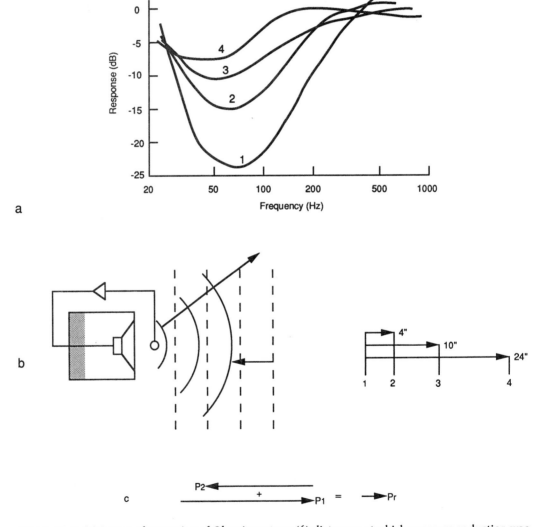

Figure 16-2. (*a*) Range of operation of Olson's system; (*b*) distances at which pressure reduction was measured; (*c*) vector relationships.

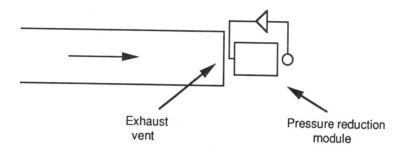

Exhaust
vent

Pressure reduction
module

Figure 16-3. Reduction of exhaust noise.

level. Complete cancellation is possible, but only at a localized point, and there are also zones where the negative vector may be dominant. As shown in figure 16-2b, the negative pressure wave extends out from the loudspeaker as a spherical wave and diminishes with distance; when the system is placed in a plane wave field, the effectiveness of the system will thus be inversely proportional to the distance from the canceling loudspeaker. The system is inherently band limited, since forcing it to work at higher or lower frequencies would, with sufficient gain, lead to instability and oscillation.

Such a system as we have described provides localized pressure cancellation, and its effectiveness is limited to a zone fairly close to it. As such, it can be used to reduce high sound pressures in certain environments where the worker's ears will be positioned consistently within the zone of system effectiveness. Wide-scale application of the technique in the workplace is not likely, since each pressure-canceling module is fairly expensive, and many of them would be required. Taking advantage of frequency and wavelength scaling, headphones have been developed that enable relatively lightweight units to function as both ear defenders and receivers for communications over a higher frequency range.

Figure 16.3 shows another application where unwanted noise can be localized; here, the pressure-reducing module is located at the output of an exhaust or ventilation system. Low-frequency sound existing at the output can be effectively canceled and not radiated further. This technique is useful when specific disturbances can be easily identified and localized.

16.3   SOUND POWER ABSORPTION AT ROOM BOUNDARIES

The basic module described in the previous section can be adapted so that it is effective in increasing the low-frequency boundary absorption in a room. Recall from section 1.7.4 that a broadband absorptive room boundary must be at least one-fourth wavelength in depth if it is to be an effective absorber. At 50 Hz, this corresponds to a depth of about 1.7 m (5.6 ft). Of course, frequency-selective Helmholtz resonators of moderate size can be constructed at the boundaries, but many of them would be needed to cover the necessary frequency range.

If the pressure-canceling module shown in figure 16-1 is modified as shown in figure 16-4, it becomes a sound power absorber. The goal in low-frequency sound power absorption is to provide a low acoustical impedance so that a given sound pressure in the vicinity of a dissipative structure will produce a fairly high air volume motion through that structure.

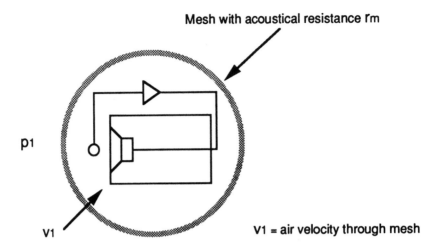

Figure 16-4. Sound power absorption.

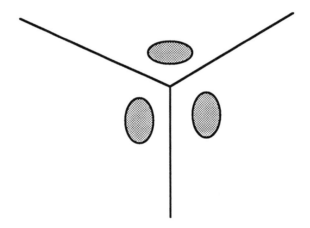

Figure 16-5. Power absorption at room boundaries.

Dissipation is provided through viscous air flow through the mesh, and the pressure-canceling module provides a low-impedance termination that allows sufficient bulk air flow to take place through the mesh. If a sufficient number of modules are located at room boundaries where the normal room modes create high pressures, then the acoustical impedance afforded by the mesh, along with the pressure-canceling module, will be quite effective in dissipating power. For best effect, the acoustical resistance r_m of the mesh should be matched to the effective acoustical free field radiation resistance.

Such absorptive elements may be built into the room boundaries, as shown in figure 16-5, or they may take the form of free-standing structures placed at or near the corners of the listening room (Pass 1988). The absorbers can approach an absorption coefficient α of unity at low frequencies, but in order to be effective, there must be a sufficient mesh area over which the action takes place. It must be clearly understood that this technique cannot make a small absorptive area look like a larger one; it can only make a given area of negligible depth appear like a broadband absorber with α approaching unity.

16.4   RECENT DEVELOPMENTS

Digital signal processing can be used in advanced noise reduction systems, as shown in figure 16-6. Here, a system is shown in which the input microphone is placed at a distance from the noise-canceling loudspeaker downstream in an exhaust duct (Erikkson et al. 1989). The reasons for doing this may have to do with extremely high noise levels and the need to reduce sounds by up to 40 dB. The IIR (infinite impulse response) adaptive filter allows for different operating parameters, such as air velocity and temperature in the duct, both of which would affect the velocity of sound from the loudspeaker back to the correction input microphone. Other reasons for locating the loudspeaker away from the input microphone may have to do with space availability and temperature constraints on the loudspeaker transducer.

The error microphone senses changes in the system's parameters and provides a correction signal for the adaptive filter.

16.5   CONCLUSIONS

While active noise control has been known for many years, it still may be considered an emerging technology. Its growth is rapid at present, and this has come about as a result of the development of microphones and loudspeakers that both are rugged and provide linear response over a wide range of operating conditions. Advanced signal processing has been an important factor here as well.

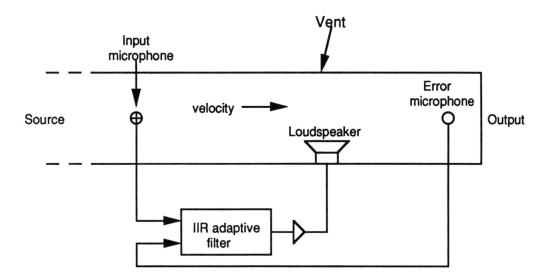

Figure 16-6. Advanced techniques in industrial noise reduction.

REFERENCES

Erikkson, L., et al. 1989. "Active Noise Control on Systems with Time-varying Sources and Parameters." *Sound and Vibration* 23(7).

Olson, H., and E. May. 1953. "Electronic Sound Absorber." J. *Acoustical Society of America* 25(6).

Pass, N. 1988. "White Paper on the 'Shadow' Room Sound Absorber." *Phantom Acoustics.*

ADDITIONAL RESOURCES

Olson, H. U.S. Patent 2,502,020.

Index